subscribed

subscribed

Why the Subscription Model
Will Be Your Company's Future—
and What to Do About It

TIEN TZUO

FOUNDER AND CEO OF ZUORA

with Gabe Weisert

PORTFOLIO/PENGUIN

Portfolio/Penguin
An imprint of Penguin Random House LLC
375 Hudson Street
New York, New York 10014

SUBSCRIBED, SUBSCRIPTION ECONOMY, and SUBSCRIPTION ECONOMY INDEX
are trademarks owned by Zuora, Inc.
Copyright © 2018 by Tien Tzuo

Most Portfolio books are available at a discount when purchased in quantity for sales promotions or corporate use. Special editions, which include personalized covers, excerpts, and corporate imprints, can be created when purchased in large quantities. For more information, please call (212) 572-2232 or e-mail specialmarkets@penguinrandomhouse.com. Your local bookstore can also assist with discounted bulk purchases using the Penguin Random House corporate Business-to-Business program. For assistance in locating a participating retailer, e-mail B2B@penguinrandomhouse.com.

Library of Congress Cataloging-in-Publication Data

Names: Tzuo, Tien, author.
Title: Subscribed : why the subscription model will be your
company's future—and what to do about it / Tien Tzuo.
Description: New York, New York : Portfolio/Penguin, [2018] |
Includes bibliographical references and index.
Identifiers: LCCN 2018010773| ISBN 9780525536468 (hardcover) |
ISBN 9780525536475 (epub)
Subjects: LCSH: Business planning—Case studies. | Strategic planning—Case studies.
Classification: LCC HD30.28 .T985 2018 | DDC 658.8/7—dc23 LC record available at
https://lccn.loc.gov/2018010773

Printed in the United States of America
10

Book design by Kelly Lipovich

While the author has made every effort to provide accurate telephone numbers, Internet addresses, and other contact information at the time of publication, neither the publisher nor the author assumes any responsibility for errors, or for changes that occur after publication. Further, the publisher does not have any control over and does not assume any responsibility for author or third-party websites or their content.

To my wife, Mariana,
for encouraging me to take the plunge

And my daughter, Ciana,
for giving it all meaning

CONTENTS

CONTENTS

PART 2
SUCCEEDING IN THE NEW SUBSCRIPTION ECONOMY

subscribed

INTRODUCTION

A couple of years ago I wrote an article for *Fortune* advising people not to go to business school. I argued that it was a waste of time, that for the past hundred years, business schools basically taught one idea: The fundamental goal of every business is to create a hit product, and then sell as many units of that product as possible, thereby diluting fixed costs in order to compete on margins. I said that this model was over, that the situation has changed.

Instead, I argued, the goal of business should be to start with the wants and needs of a particular customer base, then create a *service* that delivers ongoing value to those customers. The idea was to turn customers into *subscribers* in order to develop recurring revenue. I called the context for this change the Subscription Economy.

Oh man, did I get a lot of crap for that article. There were comments like: Do you really think we don't get it, Tien? That we don't understand the difference between a product and a service? That they don't talk about this stuff in business school? My situation wasn't helped by the fact that I still work with my old business school a lot. I go back every year to give talks and help teach courses. I got some sideways looks.

Okay, some of these people had a point. I graduated from business school in the late nineties. I'm sure some curriculum topics have changed since then. But I bet a lot still hasn't changed, particularly in those 101 classes. In fact, I know it. Every day I see companies run by bright young MBAs going down in flames trying to chase after magical hit products. They can't compete because they're built backward: product first, customer second. That order needs to flip. None of these companies has any idea who it is trying to sell to.

Here's a question: How many of the charges on your last credit card statement were made without you ever having to pull out your credit card? There are probably monthly charges on there for your Netflix and Spotify Premium accounts. There's a charge for Dropbox on there, if you keep your files in the cloud, like the savvy reader I know you are (you're reading this book, after all). Maybe you get a meal or a snack box, or have a MoviePass subscription, or support a podcast on Patreon. You're not as interested in owning stuff so much as accessing services to meet your needs.

What about your computer desktop at work? Does it still start up with the chime and the rolling green hill and half a dozen slow, cranky applications scrolled along the bottom of the screen? I sincerely hope not. It probably looks a lot simpler: a log-in, a few light desktop apps, and a browser. Maybe your company has switched over to Gmail for all its email hosting, so you don't have to delete all your old Outlook files every six months. Maybe they use Box to handle all their file storage now, and the old server room got converted into a ping-pong lounge.

Everything feels different right now. Why? Because I think

2

we're in a pivotal moment in business history, one not seen since the Industrial Revolution. Simply put, the world is moving from products to services. Subscriptions are exploding because billions of digital consumers are increasingly favoring access over ownership, but most companies are still built to sell products. They're not set up correctly for the next hundred years of business. As a result, huge opportunities are up for grabs. If you're not shifting to this business model now, chances are that in a few years you might not have any business left to shift.

WHY THIS BOOK, AND WHY NOW

Ten years ago we were already starting to see the signs. Netflix was still delivering monthly DVDs in the mail, but it was already killing Blockbuster and changing how we consumed media. Online streaming was just around the corner (as many people have pointed out, Reed Hastings called it Netflix for a reason). Zipcar was also a really interesting new concept. It was initially seen as an hourly competitor to Hertz and Budget, but you could already see new ideas opening up around cars and transportation, which Uber and Lyft capitalized on later. And of course the iPhone had just come out—at the time it was more of a fun, plug-and-play app container, but there was the potential for geolocation, identity, messaging. As bandwidth increased and platform costs decreased, there was a logical progression going on toward on-demand, digitally enabled services. And it was happening everywhere.

That's when we decided to start a new company called Zuora. We wanted to build a brand-new subscription billing and finance platform. Like a lot of companies at the time (Zendesk and customer support, Okta and passwords, Xero and bookkeeping), we were trying to solve a big, boring, pain-in-the-ass problem. To an entrepreneur, any business process that is universally hated, hopelessly complex, and massively expensive constitutes a huge opportunity. Keep in mind this was all happening in the midst of the Great Recession of the late 2000s. On-premise software got hit hard. Retail got decimated. Car sales went off a cliff. Advertising evaporated.

After the rug got pulled out from beneath them in 2008, a lot of businesses and investors realized that they were playing their own version of Hollywood economics: Pour tons of money into developing a product, then pray for a hit. If it doesn't work out, you're out of luck. These companies had no visibility into their finances. They had no predictability in their forecasts. They started every quarter with nothing in the bank and had to crawl all the way up to hit their number. That's not how it works with subscription revenue. A $10 million company with 80 percent subscription revenue starts every year with $8 million in the bank. If stock valuations are forward-looking predictions, then subscriptions are forward-looking revenue models.

None of the founders of Zuora were new to these issues. I had the good fortune to join Salesforce as employee #11, and helped turn it into a billion-dollar company over the next ten years. All of us early folks at Salesforce came out of the traditional, on-premise software industry. We were all pretty fed up with it. We thought that companies like Oracle, Siebel, and others were creating a needlessly complicated product that was sold by a mercenary sales force and promoted by a parasitic systems integration industry. The Y2K scare was in full force. Sales reps outnumbered developers ten to one. Half the installations would never see the light of day, and even those that were deemed a "success" were hated by the end users. The industry had completely lost sight of its customers: who they were, what they did on a daily basis, what they liked about their work systems, and what made them angry. It was time for a change.

Working out of Marc Benioff's rented one-bedroom apartment, we knew we wanted to build a new kind of user experience, one that would feel as seamless and intuitive as buying a book on Amazon. But as we got into it, we realized this required us to change our whole way of thinking. We had to reevaluate the whole purpose of a software company, changing the fundamental question from "How many products can I sell?" to "What does my customer want, and how can I deliver that as an intuitive service?"

When Salesforce launched, everyone realized that it was different—no more huge installations and piles of hardware. It was software as a *service*, not a static product. Because of that, we were empowered to come up with new ways to market it, to sell it, and to run a truly subscription-based company. The ideas we came up with—usage-based pricing, multitiered editions, customer success organizations—are now standard operating procedure for software-as-a-service companies, but when we were building Salesforce, none of them existed. We had to invent them.

But starting everything from scratch had its downsides, too. We knew, for instance, that to make this new business model work, we had to have a completely different set of back-office systems— something closer to a telco or a publisher, which I was familiar with from my Oracle days. But there was nothing to buy off the rack; everything was built for giant phone or energy companies, so we had to build everything ourselves. It was costing us millions of dollars every year to build out the billing, the commerce, the quoting, the whole infrastructure. We quickly realized that was a problem. But we also knew that putting a bunch of engineers on building a homegrown billing solution rather than concentrating on our core product wasn't a very good idea, either.

In 2007 Marc was talking to some guys from WebEx—K. V. Rao and Cheng Zou—and he casually invited me into the meeting. We wound up spending half the time kvetching about our billing system. Marc was agonizing over the fact that he had to spend another few million dollars building this awful homemade billing solution, and Zou said: "Oh, yeah. We have the same problem. It's a nightmare. We have a team of forty or fifty people on it." Then Rao said, "Gosh. If Salesforce and WebEx are both having this problem, maybe this is a good business to start." Maybe.

We continued the discussion over the next several months. Rao was really keen on the idea of a SaaS company dedicated to subscription billing, but I wasn't sold right away. We faced the same questions that every potential start-up faces: Who are we going to sell to? How big is this market, really? Is this going to be a software company that sells only to other software companies, or is it something bigger? And the more I thought about it, the more I realized there was nothing about the idea of subscriptions that was fundamentally limited to the software market. I also realized that all the stuff I learned at Salesforce—not just the technology, but also the innovating, marketing, and selling—would be valuable to all kinds of subscription businesses, in all kinds of industries.

Today Zuora has a thousand customers in dozens of industries. We work with streaming media companies, publishers, newspapers, manufacturers, online learning companies, health care providers. We work with giant tractor companies and small cannabis infrastructure start-ups. Our clients run planes, trains, and automobiles. We manage billions of dollars of subscription revenue every day. As a result, we know a lot about this model and how it applies to all sorts of fields. We've found, for instance, that companies running on subscription models grow their revenue more than nine times faster than the S&P 500 (check the Subscription Economy Index at the end of this book for the latest data on that topic). Depending upon the size and type of your company, our development group has done a lot of research on specific goals that you should try to hit, as well as unique threats that you should try to avoid.

WHAT YOU WILL LEARN IN THIS BOOK

I can't tell you how often, at the end of one of my presentations, someone asks me for fundamentals on how to transition a traditionally product-based company to a subscription-based revenue model. After all, competitors can steal your product features, but they can't steal the insights you gain from an active, loyal subscriber base. I assume what you're looking for is some clarity around how the model works and how to best apply it within your company. Some industry benchmarks, relevant case studies, and best practices wouldn't hurt. That's the goal of this book.

I'm also trying to fill a gap—there's a strange absence of decent resources on this topic. There's a lot of material on consumer membership programs and subscription boxes, and vast amounts of inside baseball stuff on SaaS metrics, but very little for an enterprise business reader looking for a basic playbook on making the shift to recurring revenue. While subscriptions have been getting a huge amount of press coverage lately, I'm going to give you the most important material, the stone tablets. In Part 1 of this book we'll explore how subscriptions are transforming several different industries, and in Part 2 we'll dig into more tactical, operational details about how to apply the subscription model across every aspect of your company. Here are some of the topics I'll be covering:

- How the subscription model is transforming every industry on the planet, including retail, journalism, manufacturing, media, transportation, and enterprise software

- The basic financial model and most important growth metrics for subscription businesses

- How the subscription model changes your approach to engineering, marketing, sales, finance, and IT

- The eight core growth strategies of all subscription businesses

- A customer-centric operating framework for subscription businesses

I want to note that this isn't just a Silicon Valley story, nor is this a Silicon Valley book (there are already lots of those). This is a business story. In many ways, this is a book that favors the so-called legacy companies, or "the incumbents." Because underneath all the hype of technological disruption is actually a very simple but powerful idea: companies are finally starting to understand their customers.

And when you finally discover your customers, it changes everything about your company. It affects every role. With a subscription model, suddenly your development team is spinning up new services based on usage data, as opposed in response to the loudest voices in the room. Your finance team is getting ahead of churn and test-driving new ideas. Your customer service team is proactively advising, as opposed to reacting to, tickets with scripts. Your marketing team can tie pricing to value, so they can come up with creative new packages and services. You're no longer hamstrung by back-end processes that can't adapt and scale. There are no more strictly linear, bucket brigade–style operations. Your organization is fluid but cohesive, recurring and responsive, and above all, relentlessly centered around your customer.

PART 1

THE NEW SUBSCRIPTION ECONOMY

CHAPTER 1

THE END OF AN ERA

What does digital transformation look like? Well, for starters, let's acknowledge that "digital transformation" is a really vague term. It's the kind of smart-sounding phrase that gets thrown around a lot in conferences and McKinsey reports and *Harvard Business Review* articles. The kind of expression that lots of people instinctively nod their head at, whether they know what it means or not. It could mean everything, it could mean nothing.

Let me tell you what I think it means. You've probably seen the statistic—more than half of the companies that appeared on the Fortune 500 list in the year 2000 are now gone. Poof. Vanished off the list as a result of mergers, acquisitions, bankruptcies. The life expectancy of a Fortune 500 company in 1975 was seventy-five years—today you have fifteen years to enjoy your time on the list before it's lights out. Why is this happening? Instead of dwelling on failure and looking at all the companies that went away, let's look at the companies that have stayed.

Notice how big manufacturing companies like GE and IBM that were on the first list in 1955—and are still on it today—don't talk as much about their mainframes and refrigerators and washing machines anymore? They talk about "providing digital solutions," which is an admittedly jargony way of saying that the hardware is just a means to an end. In other words, these companies now focus on achieving outcomes for their clients, rather than just selling them equipment.

GE was #4 on the first Fortune 500 list in 1955, and it's #13 on the list as I write this book in the fall of 2017. GE was incorporated as the Edison General Electric Company in 1889. It made and sold lightbulbs, electrical fixtures, and dynamos. Today GE generates most of its revenue from services, not products. GE ran commercials during the Oscars with the tagline "The digital company. That's also an industrial company." Notice the switch there. This transformation is what allows GE to survive and remain on the Fortune 500 list.

IBM was #61 on the Fortune 500 list in 1955, and it's #32 on the list today. IBM originally sold commercial scales and punch card tabulators. Today it sells IT and quantum computing services. It has completely transformed from a product manufacturer into a business services giant. IBM is now working on Watson—a technology platform that uses natural language processing and machine learning to reveal insights from large amounts of unstructured data. It has Bob Dylan chatting with an artificial intelligence system in its advertisements. It is now in the business of cognitive services—a pretty exciting departure from where the company started.

In fact, 12 percent of the companies on the 1955 Fortune 500 list are still on it today, and most of them have similarly transformed. Xerox has moved from manufacturing photographic paper and equipment to information services. McGraw-Hill has moved from printing textbooks and magazines with titles like *American Journal of Railway Appliances* to offering financial services and adaptive learning systems. NCR went from selling cash registers to saloons during the days of the Wild West to creating digital payment services that compete with companies like Square. They don't really sell *stuff* anymore.

Okay, what about some of the more recent entrants on the Fortune 500 list? The "new establishment" companies like Amazon, Google, Facebook, Apple, Netflix. The companies that instantly feel very familiar to us but are actually relatively new to the Fortune 500 list. They've rocketed to the top of the list and show no signs of going anywhere. They never thought of themselves as product companies—no transformation was needed. From the start, these companies were relentlessly focused on building direct digital relationships with their customers. And established enterprises are taking note.

Let's take a look at one big company we're all quite familiar with—Disney. Its CEO, Bob Iger, said recently, "It's one thing to be as fortunate as we are to have Disney, ABC, ESPN, Pixar, Marvel, *Star Wars* and Lucasfilm, but in today's world, it's almost not enough to have all that stuff unless you have access to your consumer." Right now, outside of its theme park attendees, Disney doesn't have much in the way of individual customer insight. Someone who buys a *Spirited Away* doll at a Walmart is a Walmart customer, not a Disney customer. Someone who goes to see a *Star Wars* movie at an AMC Theater is an AMC Theater customer, not a Disney customer. For Disney, it sounds like that's all about to change very soon.

Finally, how about the up-and-comers, the companies that may soon top the Fortune 500 list, new disrupters like Uber, Spotify, and Box? These companies came in and took everyone by storm. They haven't just gone beyond selling products, they've invented completely new markets, new services, new business models, and new technology platforms, leaving many established companies trying to play catch-up. As consumers, we love these brands, we love these services, and we love the value they provide us—a value that goes way beyond what a single product could ever offer.

What are the common threads among these three groups of companies? Whether it's GE, Amazon, or Uber, they are all succeeding because they recognized that we now live in a digital world, and in this new world, customers are different. The way people buy has changed for good. We have new expectations as consumers. We prefer outcomes over ownership. We prefer customization, not standardization. And we want constant improvement, not planned obsolescence. We want a new way to engage with business. We want services, not products. The one-size-fits-all approach isn't going to cut it anymore. And to succeed in this new digital world, companies have to transform.

THE PRODUCT ERA AND THE TYRANNY OF THE MARGIN

For the past 120 or so years, we've been living in a product economy. Companies designed, built, sold, and shipped physical things under an asset transfer model. Business was about inventory, shelving, and cost-plus pricing. The relationship between seller and buyer was based on discrete, often anonymous transactions. The sign by the cash register summed it up: "ALL SALES FINAL." Early retail pioneers like Sears and Macy's changed the way mass society consumed things, but they had minimal insight into who was actually buying their products or how they were using them.

When Henry Ford's first moving assembly line went into operation in 1913, it was really just an extension of manufacturing principles first put in place during the Industrial Revolution of the 1800s. The assembly line wasn't just about maximizing efficiency through discrete repetitive tasks, it was a metaphor for how a company's product can dictate its supply chains, manufacturing processes, distribution channels, and management layer.

The product was the only governing principle—it organized everything across a perfectly straight line. The actual people involved in making, buying, and selling the product were entirely disposable. Henry Ford's customers could famously pick any Model T color they wanted, as long as it was black. The result of all this relentless efficiency was that Henry Ford's cost per unit dropped precipitously, allowing him to flood the market with cheap but durably made cars. Model Ts came only in black because with one automobile coming off the line every three minutes, *that was the only color that would dry fast enough.*

Then once these big companies established market share, the thinking went, they could start to gently raise their prices and make money off the difference, or margin. The margin ruled everything (and a little planned obsolescence never hurt). It's difficult to overstate the power that big postwar American corporations had. They organized themselves around strictly delineated product divisions and didn't have to answer to anyone. There were no call centers, no customer service reps, and, in many cases, no returns, period. This model didn't work particularly well when it came to customers like our grandparents, but it consistently shipped units and kept boardrooms happy.

The emergence of enterprise resource planning (ERP) systems in the latter half of the century only exacerbated this problem. These systems did a good job of measuring operational efficiency: raw materials, inventory, purchase orders, shipping, payroll. They did a lousy job of measuring actual customer experience. But as modern management guru Peter Drucker pointed out, companies tend to manage what they can measure, and so executive teams became hopelessly product-focused, both organizationally and strategically.

This period also saw the ascendance of supply chain economics. The goal was to match supply and demand with the least inventory possible. It was nirvana for engineers and management consultants, who were threatened by the new electronic products and efficiencies coming out of Japan. "Just in time inventory" meant that warehouses full of stuff just sitting around were the ultimate enemy. "Total quality initiative" meant that the work of improving processes was never over. Michael Dell built an empire based around this discipline.

Then around twenty years ago, corporate America woke up to the realization that all this relentless focus on productivity was coming at a cost—namely the relationship between the vendor and the customer. The customer was a complete unknown, a receptacle at the end of a distribution chain whose sole purpose was to "consume" the products companies made. And as it turned out, many of these new consumers were having difficulty getting their new products to work. And how did corporate America discover this? Their receptionists were getting angry phone calls.

So what did the big companies do to address this problem? They set up customer service departments! When in doubt, build another vertical silo—they launched market services, technical support lines, warranty contracts, and maintenance groups. The customer had truly arrived—they had their own department now. And that department was located way down at the far end of the supply chain, just past the loading dock.

THE AGE OF THE CUSTOMER

Today the glory days of the soulless, all-powerful corporation are long gone. Today's customers are more informed by an order of magnitude. Most of them have researched, assessed, and categorized you before you can even say hello. And to most of them, especially younger ones, ownership just isn't that important anymore. People increasingly view the prospect of buying something as unnecessary baggage. They want media at their fingertips, not physical products to manage. That's why most of the big box retailers I grew up with are gone now: Circuit City, Tower Records, Blockbuster, Borders, Virgin Megastore. A lot of the malls are gone, too! Today people expect services to provide immediate, ongoing fulfillment, from rideshares to streaming services to subscription boxes. They want to be happily surprised on a regular basis. And if you don't meet those expectations, you get dropped, not to mention trashed on social media. It's that simple.

Forrester Research thinks we're at the beginning of a new twenty-year business cycle it calls "The Age of the Customer." Forrester sees a broad, systemic shift in capital models pivoting toward serving a newly empowered generation of customers who have the ability to price, critique, and purchase anytime, anywhere. Here's how Forrester describes the new customer mindset: "The expectation that any desired information or service is available, on any appropriate device, in context, at your moment of need." Customers have new expectations (and yes, those expectations have certainly been driven by millennials, but at this point, almost everyone shares them). They want the ride, not the car. The milk, not the cow. The new Kanye music, not the new Kanye record.

Initially, the corporate world responded to this shift in pretty typical fashion—they built more systems. They spun up customer relationship management (CRM) databases, installed customer loyalty programs, offered membership rewards and incentives, and showered people with customer satisfaction surveys. It was a truth universally acknowledged that new customers were harder to acquire than it was to retain loyal ones, and negative customer experiences traveled much further than positive ones. There was a lot of talk about customer journeys and net promoter scores.

No one knows who coined the phrase "the customer is always right," but it dates to late-nineteenth-century department-store pioneers like Harry Gordon Selfridge and Marshall Field. It was a novel concept at the time (displacing a prevailing general retail attitude of caveat emptor), but what's amazing is how all these big Fortune 500 companies still couldn't get it right. They developed a lot of prescriptive strategies built around customer focus, but they lacked a *descriptive* understanding of the mindset of the customer herself. Large companies were still getting blown up on social media left and right, and there were certainly no sweeping changes in public sentiment toward big enterprises. It just wasn't enough.

And then a funny thing happened—those digital disrupters like Salesforce and Amazon that I mentioned earlier took the whole customer-first concept a huge step further by *actually establishing direct ongoing relationships with their customers*. They didn't have customer segments anymore—they had individual subscribers. And every one of those individual subscribers had their own home page, their own activity history, their own red flags, their own algorithmically derived suggestions, their own unique experiences. And thanks to subscriber IDs, all the boring transactional point-of-sale processes disappeared. Ten years ago there was no Spotify, and Netflix was a DVD company. Today both those companies own a significant percentage of the total revenue of their respective industries! Today businesses are asking themselves a whole new set of questions: What do we need to do to build long-term relationships? What do we need to do to focus on outcomes and not ownership? To invent new business models? To grow our recurring revenue, and to deliver ongoing value?

So again—what does digital transformation look like? I think it looks a lot like a circle. Let me explain.

THE NEW BUSINESS MODEL

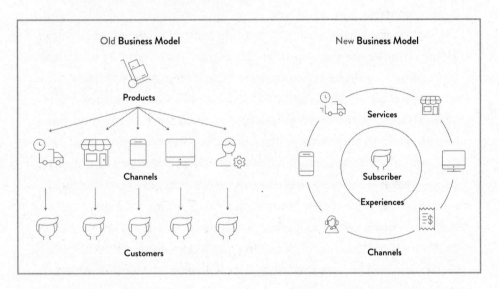

If you remember one thing about this book, remember this diagram. It summarizes the shift under way. On the left side, you have the old model, where companies used to focus on "getting a product to market" and selling as many units of that product as possible: more cars, more pens, more razors, more laptops. They did this by getting their products into as many sales and distribution channels as possible. Of course there must be a customer on the other end buying all this stuff, but often you didn't really care who they were, as long as more units flew off the shelves.

That's not how the modern company thinks. Today successful companies start with the customer. They recognize that customers spend their time across many channels, and wherever those customers are, that's where they should be meeting their customers' needs. And the more information you can learn about the customer, the better you can serve their needs, and the more valuable the relationship becomes. That's digital transformation: from linear transactional channels to a circular, dynamic relationship with your subscriber.

Big changes are coming. If you don't find out who your customers are in the next five to ten years, you will fail. Smaller start-ups are taking down huge enterprises simply because *they know who they are selling to*. The entire $80 trillion economy is up for grabs. Companies that survive over a long period of time follow their customers; they do not expect customers to follow them. Companies that know what their customers want, and how they want it, will succeed over companies that spend a lot of time and effort creating a product *they think* is a good idea, then spend equal amounts of time and effort trying to persuade people to buy it.

This shift, from a product-centric to a customer-centric organizational mindset, is a defining characteristic of the Subscription Economy. Today the whole world runs "as a service": transportation, education, media, health care, connected devices, retail, industry. Subscriptions themselves, of course, aren't new. The most basic definition of a subscription is simply a piece of writing beneath a document (sub script): a name, a note, an addendum. When two parties are involved, that constitutes a mutual agreement, an accord, a relationship. As a business model, subscriptions have been keeping journalists, authors, illustrators, historians, and cartographers paid for hundreds of years. Subscriptions also sold a lot of bad CDs in the eighties (more on that later).

So why is this shift happening right now? Because of the way those subscriptions are being delivered—digitally—and the huge amount of data those digital subscriptions are generating. Considering that business is still governed by bookkeeping standards developed in the fifteenth century, the commercial internet is relatively new—roughly only twenty years old. I grew up completely without it, and I'm not *that* old. The iPhone is just over ten years old—think of the way that device has shaped how we use services. The cloud has profoundly reshaped the ways that companies think about IT infrastructure, professional services, and capex versus opex spending. This whole new world of connected devices definitely feels very new. And as Mary Meeker noted in her latest Internet Trends report, digital consumer subscriptions are exploding because of massive new improvements in digital user experiences, particularly for mobile phones.

It feels like we're at the beginning of something very big.

So let's explore some of the ways the subscription model is transforming every sector of the modern economy.

CHAPTER 2

―――――――

FLIPPING THE RETAIL SCRIPT

Brick and mortar retail is dying. Or at least that's what the data suggests. In the United States, more stores closed in 2017 than any other year on record—at least 7,000 brick and mortar stores were shuttered, beating the record of just over 6,000 set during the financial crisis of 2008. That number represents more than five square miles of empty retail space. The names are all familiar: Staples, Kmart, JCPenney, Sears (in the 1960s, sales at Sears accounted for 1 percent of US GDP). At least a dozen heavily leveraged retailers filed for bankruptcy in 2017. As a country, we are way overstored—private equity companies loaded up the big box retailers with debt and forced them to open hundreds of new outlets in questionable locations. Roughly a quarter of the more than 1,000 enclosed malls in the United States are expected to close within the next five years (the total number of malls peaked at around 1,500 in the nineties). A quick retail industry Google search turns up "dead mall" fan sites with names like "Label Scar" for the faded markings left behind when sign lettering is removed.

Ecommerce, the thinking in Silicon Valley goes, is clearly the future. It now represents more than 13 percent of the total retail market and is growing at 15 percent a year, versus just 3 percent for brick and mortar retail. As of this writing, that ecommerce dollar amount currently stands at around $450 billion and is expected to surpass $500 billion by the end of 2018. In the United States, for example, Amazon now has more than 90 million Prime members, or roughly half of all American households. Those customers pay almost $9 billion in membership fees alone and spend an average of $117 billion a year. Online subscription programs for household staples and routine purchases are exploding in popularity, and in a few years online grocery delivery seems poised to become the next new normal.

But wait—not so fast! Over 85 percent of all US retail sales still happen in physical stores, representing more than $5 trillion in total sales. And remember, that number is still going up. Over the next four years, the global retail sector will add $5 trillion in sales for a total of $28 trillion, and most of that $5 trillion will go to physical stores. And something else interesting is happening—online brands are opening stores. Lots of them. As I write, companies like Trunk Club (clothes), Warby Parker (glasses), UNTUCKit (shirts), Casper (mattresses), Birchbox (cosmetics), Allbirds (shoes), Boll & Branch (sheets), Away (luggage), ModCloth (clothes), and Rent the Runway (clothes) are in the midst of opening *hundreds* of new physical locations. According to real estate data company CoStar Group, the square footage of physical stores occupied by retailers that started online has increased tenfold over the last five years. What's even more interesting—companies are starting to pull products off their ecommerce sites in order to drive traffic to their stores. For example, you can't actually buy coffee on Starbucks.com anymore.

And let's not forget Amazon, which now owns 460 Whole Foods stores in addition to several new bookstores and is now breaking out revenues from physical stores in its quarterly reports. According to its own internal documents, Amazon sees potential for at least 1,500 more grocery stores. As Reid Greenberg, lead researcher at Kantar Retail, says: "It isn't that retail is dead. Roughly 85–90 percent of retail takes place in brick-and-mortar locations. But bad brick-and-mortar is. These mall-type department stores are faced with many challenges because they aren't connecting with shoppers in the way they want to be connected with. Consumers already know what to expect when walking into one of these stores."

And as it turns out, it's increasingly difficult to establish yourself as a pure-play ecommerce vendor these days. According to Michael Wolf's latest *Activate* study, the top fifteen ecommerce marketplaces in the world account for more than 60 percent of total sales. As RetailNext CEO Alexei Agratchev notes, "As an ecommerce vendor, you have really high variable costs around shipping and returns. On the other hand, Amazon is an amazing logistical machine, and they're not even running at a profit most of the time. Also, customer acquisition costs are going up online—more networks are asking for more money for referrals. And the other question is—how do you really differentiate yourself online? Anything you do on your website, a competitor can steal pretty easily. But you can actually create really cool experiences in stores. Experiences that can't be found anywhere else."

IT'S NEVER BEEN ABOUT ECOMMERCE

As it turns out, stores are still incredibly valuable, and brick and mortar retail is far from dead—traditional retail just needs to flip the script. What do I mean by that? Let's start with an interesting question: What was the first thing you ever bought on Amazon?

It's sitting right there, on the last page of your order history. Go ahead, put this book down and look it up. Here, I'll show you mine:

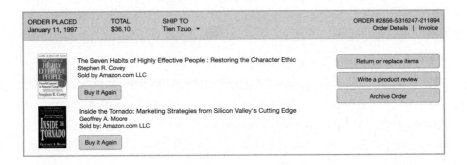

This was back when Amazon used to just sell books! Like a good business student, I spent $36.10 on Stephen R. Covey's *The 7 Habits of Highly Effective People* and Geoffrey A. Moore's *Inside the Tornado: Marketing Strategies from Silicon Valley's Cutting Edge*. Of course, I can also see everything else I ever ordered on Amazon. And as a result, Amazon knows what I'm interested in, offers smart recommendations, and takes care of the details. Everything tracks back to the customer.

Now let's take a look at Walmart. Ninety percent of all Americans live within twenty minutes of a Walmart store. Walmart has 5,000 stores, more than 1.5 million employees, and serves more than 140 million shoppers a week. Nearly every American spent money at a Walmart last year. Those are pretty amazing statistics. This is a company with decades of institutional experience with supply chains, transport logistics, inventory management. It knows how to buy and sell products. Most of its customers shop for groceries and basics—very simple, repeat purchases. But what was the last thing you bought at Walmart? They certainly couldn't tell you. To Walmart, you're basically just a vehicle for dispensing inventory. Once you pass the cash register, you vanish off the map.

To be fair, things are changing for Walmart—it has invested heavily in ecommerce, payment apps, and pickup and delivery services. But for a long time, Walmart thought like a product company. Its stores existed to sell products. Its customers were simply there to buy products. And that's just not how Amazon thinks. "We've had three big ideas at Amazon that we've stuck with over the years," said Jeff Bezos. "Put the customer first. Invent. And be patient." Another favorite Bezos quote: "I don't know about you, but most of my exchanges with cashiers are not that meaningful." The Amazon versus Walmart battle has been framed as ecommerce versus traditional retail, but that's always been a false dichotomy. It's about starting with the customer instead of the product. It's about establishing ongoing relationships. It's about flipping the script—starting with the digital experience, and then building the store.

APPLE AS A SERVICE

Every year, several thousand tech journalists show up to hear Apple executives wax enthusiastic about a new iPhone that is a little bit thinner, shinier, and wider than the iPhone they talked enthusiastically about last year. As transformative as the first iPhone was, today this whole exercise is getting to be pretty silly. As Goldman Sachs analyst Simona Jankowski rightly notes, "The smartphone battleground is shifting from unit land grab to user monetization." I think the Apple executives must realize this as well, because lately they've been taking great pains to point to the growth of their service revenue, which they hope to double by the year 2020. Its February 1, 2018, earnings call was almost exclusively dedicated to highlighting its service revenue, which was

$31.15 billion in 2017 and could constitute a Fortune 100 company itself. That revenue is growing at 27 percent a year and represents more than half of Apple's growth. And while its hardware business is seasonal and subject to wide peaks and troughs, its service business shows consistent, predictable growth quarter over quarter. But guess what? Some people still don't get it! The Q&A session of that last earnings call was dominated by analyst questions around iPhone supply and demand. It's enough to make you slam your forehead on your desk.

Now, I understand that today Apple is doing just fine by selling expensive phones to affluent people. But just imagine what would happen at the next Apple keynote if Tim Cook announced a simple monthly Apple subscription plan that covered *everything*: network provider charges, automatic hardware upgrades, and add-on options for extra devices, music and video content, specialty software, gaming, etc. Not just an upgrade program, but Apple as a Service. I have to admit, this isn't my idea—it belongs to Goldman Sachs. Goldman analyst Jankowski has suggested an "Apple Prime" $50 monthly subscription that would include a guaranteed phone upgrade, Apple TV, and Apple Music. If, as a result, Cupertino could deliver a financial statement similar to Salesforce's noting that 80 percent of next year's revenue was already in the bank, how fast do you think it would take for Apple to hit that trillion-dollar valuation?

As each year passes, Apple cares less and less about how many iPhones it ships, and more about its revenue per Apple ID, lifetime value per Apple ID, and efficiency metrics toward growing the base and value of those Apple IDs. Apple has cleverly integrated those IDs into its retail experience as well. I can walk into any Apple store, give them my ID, and walk out with a product. That's pretty amazing. Starbucks also has the ID. I can log in to Starbucks and look at all the coffees and lattes I've been drinking since I started using a Starbucks card and its mobile payment app. Who else has the ID? There aren't too many companies that you can name off the top of your head. That's a sobering lesson for any consumer brand.

THE NEW BOOK OF THE MONTH CLUBS

If you're still selling your product off shelves to strangers in five years, there's a good chance you're not going to make it to ten. Today the overwhelming imperative of every consumer brand should be to *know your customer*. If you don't do that, you will fail—plain and simple. That's what all these new phenomenally successful subscription box companies understand: Birchbox (cosmetics), Dollar Shave Club, Loot Crate (stuff for gamers), Stitch Fix (clothing), Freshly (meals), Graze (snacks), Trunk Club (clothing), Fabletics (active wear), Stance (socks). All of these companies are very different from generic "X of the month" clubs. They have taken a customer-first approach to retail by starting directly with the customer and creating fun, compelling subscription experiences that get smarter and smarter over time.

It wasn't always this way. Columbia House was a mail-order music service that holds a special place (dreaded place?) in the hearts of many of us who grew up in the eighties and nineties. Sign up for an introductory deal and, for a mere penny, Columbia House would send you the albums (and later, eight-tracks, cassettes, and CDs) of your choice. Journey? Run-DMC? Springsteen? It didn't matter: Columbia House wouldn't judge. You just made your picks and then awaited the mailman's delivery. But after decades of dominating the mail-order music space—pulling in as much as $1.4 billion in sales in its heyday—Columbia House filed for Chapter 11 bankruptcy late in 2015. While some of the blame for Columbia House's dissolution can be placed on our changing trends in media acquisition (read: Spotify, Netflix, niche OTT media providers, and more) and stiff competition from Amazon and other online retailers, I think that much of its collapse stemmed from mismanagement of subscriber relationships.

Rather than trying to build a business around its relationship with its subscribers, Columbia House shipped products by default and made consumers pay the price. It's a sad trick—consumers get sucked into providing their credit card information for a discounted product or free trial and then, when they fail to cancel in time, they continue to get billed. Thanks to unclear billing practices, difficult cancellation processes, and poor communication, lots of subscribers got stuck with junk they didn't want. Sadly, lots of companies still depend on customer neglect in order to sustain their zombie business models.

So what's changed about the monthly subscription services today? Well, the smart ones realize that if they really want to retain their subscribers, they need to focus on building a great service, without relying on lame tricks like hiding the cancel button. "If you are in the subscription business for the long term, especially in an age of lightning-fast communications among dissatisfied customers, you want to follow the golden rule," says Robbie Kellman Baxter, author of *The Membership Economy*. "Make it easy for customers to leave if they want to. You can certainly ask them why they're leaving, or try to win them back, but don't get in their way—the digital equivalent of blocking the exit with a hulking security guard."

We've also seen companies like Fabletics (a Lululemon competitor fronted by actress Kate Hudson) and Adore Me (designer lingerie) come under fire for poor management of their subscription businesses. Unclear billing practices, difficult cancellation processes, poor communication—these subscription retail issues (and customer complaints) all flow from this idea of the "negative option" model. To their credit, both companies took concrete steps to improve service and transparency, raising their customer satisfaction scores in the process. But more than sixty years after Columbia House was started, it seems that many other subscription retail businesses still don't get this point. They have no idea how to sell a great customer experience—not just a product—and how to capitalize on the potential for a seamless customer experience. You can't just slap a monthly fee on a product and start shipping—you have to change your entire way of thinking. You need a mindset that treats your customers like subscribers—partners in an ongoing, mutually beneficial relationship.

FENDER: FROM SELLING GUITARS TO CREATING MUSICIANS

Subscriptions are obviously great for perishables and repeat purchases: razors, diapers, groceries, detergents, pet food. But what if you sell more expensive products but still want to take advantage of the kind of subscriber relationships that those monthly box services enjoy? Well, you wrap compelling digital services around them. Fender is a great story here. They've been making amazing electric guitars for more than seventy years. But sales of electric guitars, industry-wide, have fallen by about a third in the last decade. And while almost half of Fender's sales are to brand-new guitarists, about 90 percent of them quit the instrument within a year. From a subscription business's point of view, that's a 90 percent churn rate!

    That's largely because guitar is "a hard instrument to learn," explains Ethan Kaplan, chief product officer–digital for Fender. It may be pretty easy to pick up a few chords, but most beginning guitar players tend to plateau after that, eventually abandoning the instrument entirely. But if Fender can keep people playing and reduce that churn, Kaplan knows that most of them will remain customers for life. Cutting abandonment rate became a key priority, and solving for that meant thinking beyond a "connected guitar." So Fender launched a new subscription-based online video teaching service called Fender Play, which teaches guitarists to perform their first riff or song in a half hour or less. (I'm a fan—so far I've learned three open string chords: C, D, and G. Let's hope I don't plateau.)

"We did a segmentation study to really understand our audience, and we used that as the nexus point to develop the digital strategy," Kaplan explains. The success of Lynda.com, the subscription-based training website, also helped convince Fender there was a market for Play's premium content. Fender Tune, a free mobile app for guitar tuning, was Fender Digital's first product offering in August 2016. Tune helped clear the way for Play—and helped Fender get up to speed on harnessing vast amounts of consumer data. "I can see minutes spent [on Tune], how many people are tuning, what they're tuning, if they're successful," Kaplan says. Before launching Play, Kaplan's team spent a year building its data analytics and dashboards for real-time insight across Fender's digital products. "Having a continual dialogue with our customers through learning is really key," Kaplan says. "I don't want to just sell people guitars and then hope they play it."

Fender CEO Andy Mooney, who incidentally closed our 2017 *Subscribed* conference in San Francisco with an amazing cover of "Layla," says that by simply reducing his abandonment rate by 10 percent, he could double the size of his market. That's a really compelling example of someone applying a service-oriented mindset to an ostensibly "static" product. Instead of thinking about reseller margins and unit sales, Mooney is thinking about subscriber bases and engagement rates. It's not about owning a guitar, it's about being a guitar player and a music lover for life. "Leo Fender actually never played guitar," said Mooney. "But he listened to artists. At Fender we still believe in listening to our customers."

THE NEW RETAIL EXPERIENCE

We've seen that retail can work, as long as you flip the script. What does this mean? Well, if you look at Amazon and Apple, everything starts with the customer (yes, I get it, Apple is famously a "product" company, but my argument is that it is increasingly—and correctly—thinking of those products as a means of enabling customer-based services). And now a whole new breed of subscription box companies is thriving, thanks to this same operational mode of relentless customer focus. Even luxury goods companies like Fender are innovating with ongoing digital relationships prioritizing learning and engagement. Now, how does this all relate to "regular" stores in malls? Again, the new breed of retailer *all starts with the customer.* Mike Elgan, a columnist for *Computerworld*, sums it up well: "The bottom line is that there is no 'retail apocalypse.' That's based on an obsolete dichotomy between 'online' and 'physical' retail. The real division is between data-driven, app-centric, flexible and omnichannel retailing on the one hand, and old and stale retailing on the other." Let's take a closer look at how companies are treating retail stores as extensions of their online stores, as it is becoming increasingly obvious that it's very difficult to be profitable through online sales alone.

Do you remember the great "showrooming" scare? Retailers used to be terrified of people coming into their stores, browsing around, then buying cheaper versions from competitors online.

After a little market research, of course, they found out that the opposite was true—more people research online first, then head to stores to try out products before they buy them. Today, that's the key to any successful consumer brand experience—the online experience has to come first. Why did Gillette's share of the US men's razor business fall to 54 percent in 2016 from 70 percent in 2010? Because more men are buying online from Harry's and Dollar Shave Club. And moving the customer identity and the shipping and packaging logistics online lets you design your physical stores as showrooms, not warehouses. Warby Parker estimates that three out of every four shoppers in its physical stores visit its website first.

The Bonobos "Guideshops" (which now belong to Walmart!), for example, don't really sell anything. If you like something in the store, they'll ship it to you later. The main idea is for people to try things on and get advice. They're using their stores to surface the discovery process, not manage inventory. Warby Parker is averaging $3,000 per square foot of retail space (slightly under Tiffany's number) by knowing that 85 percent of their foot traffic has already done extensive browsing online. They don't try to clutter every inch of their retail space with stuff to buy. Nordstrom's new local stores feature all kinds of services like styling and manicures for their Trunk Club members, but they don't feature any actual inventory. Customers get their goods through curbside pickup or same-day delivery. Amazon's new bookstore seems to be displaying its books (gasp!) face-out, with accompanying commentary and ratings cards. This is a lot more consumer-friendly than throwing shoppers at a wall of shelved books. They're surfacing new and interesting content, the same way Netflix does on its home page. As Starbucks CEO Kevin Johnson explains: "Number one, you must be focused on experiential retail that creates an experience in your store that becomes a destination for the customer. And number two, you have to extend that experience from brick-and-mortar to a digital-mobile relationship."

Naturally, all of these companies are taking advantage of online data to inform the design and presentation of their physical stores. In its New York store, Birchbox uses rankings and reviews from its website to inform the way it arranges its physical inventory. Birchbox also does simple, intuitive things, like organizing by category, not brand. Tesla dealerships aren't huge lots of cars swarming with salespeople working on commission. The point is to inform and answer questions. If you like the car, you can take care of the transactional stuff online.

Pickup and subscription delivery services are increasingly becoming the norm. According to McKinsey, the subscription ecommerce market has grown by more than 100 percent a year for the past five years. Particularly for things like groceries and staples, some sort of expedited pickup or delivery is the new table stakes. Retailers are working with car companies to develop trunk locks with independent cryptosecurity features so that third parties can deliver dry cleaning or groceries to your car while you're at work. Target is realizing that Amazon has around eighty fulfillment centers in the United States, while it has potentially eighteen hundred. Like Walgreens, today Target is driving a ton of in-store pickup traffic through its app.

Even the basic economic model of a retail store is being reinvented in all sorts of cool ways. b8ta, a retail store that sells trendy tech gadgets, *doesn't make any money from product sales*. Its business model is entirely based on paid subscriptions from the product vendors themselves, which keeps it laser focused on boosting their return on investment. What's more, that steady, recurring revenue model makes b8ta far less dependent upon nailing its fourth-quarter holiday sales in order to prop up the rest of the year. Win-win. And those dying malls? Well, the ones that aren't going under are doing quite well. As Alexei Agratchev of Retail-Next notes, "As old stores that average $250 sales per square foot are replaced by hip new brands that are averaging $700 to $800, they optimize the overall portfolio so that successful malls are doing better and better." The thriving Westfield World Trade Center mall in downtown Manhattan, for example, is part museum, part entertainment complex, part showroom, part social watering hole. Its stores are doing fine, but management understands it needs to invest in access and amenities to keep feeding the Disney Parks "flywheel" growth model of foot traffic, entertainment value, and dining and retail sales.

HUSQVARNA, THE 329-YEAR-OLD START-UP

Finally, what if you don't have a store—you have a shed? Another great story of a retailer surfacing value through subscription revenue and convenience is Husqvarna, which is a national institution in Sweden—it was founded as a weapons foundry in 1689. Today, it's a world-leading manufacturer of products for forestry and lawn and garden maintenance, as well as tools for the construction and stone industries. For its core home consumers, it has recently rolled out something called the "Husqvarna Battery Box." It looks kind of like a fancy blue storage shed that's located in the parking lot of a popular shopping center.

It's actually a commercial tool library. Husqvarna subscribers in Stockholm can take advantage of the Battery Box to access all kinds of heavy, battery-powered equipment like hedge trimmers, chainsaws, and leaf blowers. The tools are serviced daily to ensure that they are always in good condition and fully charged before customers take them home. Subscribers pay a flat monthly fee and simply return stuff when they're done—no storage, no maintenance, no hassle. It's also a great opportunity for people to try out tools before purchase. "People are already sharing homes and cars. To share products that are only used occasionally, like a hedge trimmer, makes a lot of sense for some users," says Pavel Hajman, president of Husqvarna.

The new business model imperative is to grow and develop a dedicated subscriber base. In the retail sector, that translates to establishing a digital identity with your customer that encourages discovery and engagement, as well as the delight of a compelling physical retail experience. The new winners are using their *physical stores as extensions of their online experiences*, not the other way around. They're flipping the script.

CHAPTER 3

THE NEW GOLDEN AGE OF MEDIA

During the Golden Age of Hollywood, from the late 1920s to the early 1960s, the "Big Five" motion picture studios cranked out dozens of films a week. The emphasis was certainly on quantity over quality, but every now and then an ordinary-seeming genre film eventually did quite well: *The Searchers, The Wizard of Oz, Casablanca*. It's important to note that these masterpieces were mass-produced along with hundreds of other westerns, musicals, and mysteries. There was a portfolio effect—the hits reliably paid for the misses. Of course this all happened when box office revenues were much more stable than they are today. For most Americans, going to the movies was a dependably recurring activity that happened on a weekly or even daily basis. The movie studios ran a relatively predictable business.

But then along came television, and suddenly the stakes became much higher. People weren't going to the theaters as much.

So how did the Hollywood studios respond? With Charlton Heston. They spent big money on epics like *Ben-Hur, The Ten Commandments*, and *Antony and Cleopatra*. They doubled down on lavish sets, casts of thousands, and bankable stars. Of course, that only got them so far—the film industry was in pretty bad shape by the mid-sixties. But then a bunch of hip new producers and directors took the old format in all sorts of interesting new directions, and by the time *Jaws* and *Star Wars* rolled around, the blockbuster was the firmly established Hollywood business model: Score the big hit, then make money off residuals like TV licensing and action figures and novelizations and Halloween costumes and candy tie-ins. Expanding overseas markets and the DVD boom only helped to lock in this product mindset of placing fewer but larger bets on big, splashy "tentpole" movies. There were lots of ways to sell a hit.

The music business essentially ran the same operation, albeit with lots more bets. Columbia Records introduced the twelve-inch vinyl album in 1948 (before that we had wax cylinders and 78s). The breakout hits paid for all the flops; Bruce Springsteen's first two records were duds before he made it big with *Born to Run*. A hit song on the radio translated into album sales, karaoke licensing, a Muzak version, an appearance on a movie soundtrack, maybe a greatest hits package. Today in Silicon Valley jargon we might call this approach "monetizing longtail content." Then when CDs arrived in the late eighties, the record industry got a chance to sell its entire back catalog again (!), as well as enjoy ridiculous profit margins on new releases. Mariah Carey's *Merry Christmas* album is a good example—it was definitely a hit when it came out in 1994, but then it just refused to stop selling (it currently stands at more than 15 million copies worldwide), and Carey continues to dependably promote it every holiday season. So, hits and misses.

Until the music stopped.

THE IMPLOSION AND THE REBOUND

Then along came the internet and file-sharing sites, and the entertainment industry responded calmly and methodically, in a studied effort to create its own legal online alternatives. Just kidding, they freaked out—there were lawsuits, congressional inquiries, kids getting hauled into court, and Metallica's Lars Ulrich delivering the names of 335,000 nefarious copyright infringers (who I assume were also Metallica fans?) to Napster's office. Then Steve Jobs came to the rescue of the music industry, sort of. You didn't even have to buy a whole album in order to listen to one or two hits—you could just buy the songs, for a dollar a pop! The labels signed up in droves, because they recognized a very familiar business model—keep pushing the hits. Ultimately, of course, the dollar-a-song model wouldn't last. It temporarily stemmed piracy and gave the labels some short-term relief, but it perpetuated the old system of relentlessly pushing Top 40 songs. Plus, the overall industry revenue kept declining, and eventually, once bandwidth caught up, the TV and movie studios started to suffer from the Pirate Bays of the world.

No one noticed at the time, but a few smart, canny start-ups started making it actually easier to consume online media legally through streaming and simple monthly subscriptions. Netflix, which started streaming movies in 2007, went from zero to 100 million streaming subscribers in ten years. Spotify, founded almost ten years later, went from zero to more than 50 million paying subscribers in less than nine years, and today is responsible for more than 20 percent of global music industry revenues. The streaming sites eventually won the online piracy war and provided a much more reliable business model as well.

And here we are today. No more trying to make ugly DVD racks look good in your living room, no more scratched CDs, no more racing home from work to catch your favorite prime-time show. That's so . . . five years ago. I grew up in the eighties, when you heard a song on the radio, went to the mall to buy the tape with the song on it (for fifteen bucks, which was a lot of money), and ran home to listen to the hit. If the whole album was good (*The Joshua Tree*), you got lucky. If it wasn't (*No Jacket Required*), you didn't. Either way, your $15 tape was going to sound like toilet paper in six months. Today algorithms and Spotify playlists have created a whole new discovery layer of music. Back then, the discovery layer was the cool kid who knew about R.E.M. and the Smiths. If you didn't have that kid in your class, you were out of luck.

Today we seem to be in a new golden age of media—one that feels similar in many ways to the heyday of the old studio system. By and large, artists still need to be paid more, but there is so much more music to explore, so many new movies and shows to discover. Freed from a blockbuster mindset, the new streaming services don't have to worry about chasing lowest common denominator entertainment. They don't have to worry about losing that minivan commercial. They can take risks on smarter, edgier projects. Could you imagine seeing *Stranger Things* or *Transparent* or *Orange Is the New Black* on prime-time television? Netflix now spends $8 billion a year on original content. That's a huge number that continues to drive the company's critics in the analyst community absolutely nuts: that's not sustainable, it's on track to getting bought in three or four years, Reed Hastings is arrogant and out of touch, etc. Unless it figures out a way to partner or work some production magic, this effort is going to cost Netflix way too much money to be successful, they say. Leave the Hollywood stuff to the Hollywood people and stick to the Silicon Valley thing of "leveraging platforms."

So what is the management team at Netflix thinking? Well, the first clue is that Netflix has 120 million subscribers around the world today. If you just multiply that by a hundred bucks a year, on average, that's $12 billion of revenue. They're obviously trying to plow as much of that revenue back into original content, but how are they going to justify that spend? The business model for an awful movie like *Batman v Superman: Dawn of Justice* is pretty straightforward—the studio spends $250 million to produce the movie, then they throw it out there, and if it generates three or four times that, then it's considered a win. And if nobody goes, it's a bust. Hollywood is still Hollywood, after all (Steven Spielberg has gone on record lamenting the blockbuster monster he helped create, predicting a future where we'll have to pay $40 for *Iron Man* but $7 for the new *Lincoln*). You spend money to create a movie, and the market either rewards you or doesn't (incidentally I want my money back for *Batman v Superman*).

The business model for a new Netflix show is fundamentally more stable. It spends about $50 to $60 million on a new season of *GLOW* or *Godless*. So how does Netflix justify its spending on a TV show or movie that it doesn't "sell"? Again, let's return to that portfolio effect. Regardless of whether a show is successful or not, investing in sharp new content helps Netflix to both (a) attract new subscribers and (b) extend the lifetime of its current subscribers. Those shows don't go away! Together, they're increasing the overall value of the portfolio. They are instrumental in driving down customer acquisition costs (as more subscribers sign up) and increasing subscriber lifetime value (as more subscribers stick around for longer). Netflix knows exactly how long it takes for a subscriber to flip from unprofitable to profitable. Spending tons of money on new shows means Netflix is happy to take a hit on the books in the short term in order to increase their profitability in the long run.

But we'll get into the wonkier stuff later. Let's look at some cool new media stories and the lessons they have to share.

NO LONGER SO NICHE: CRUNCHYROLL AND DAZN

Roughly two-thirds of all Americans now subscribe to a streaming video service. And if it's not screamingly obvious already, every video content provider on the planet, from the biggest national network to the tiniest cable channel, is transitioning to SVOD, or subscription video on demand. There are streaming media services for every genre under the sun: Bollywood films, British comedies, Korean soap operas. Livestreaming is another huge growth area—look at the explosive popularity of the video game streaming site Twitch, which now attracts almost a million viewers a month. The steady and predictable income generated by subscriptions lets these companies make considered investments toward their future (whether sourcing great backlist content or developing original material) rather than scrambling to hit their numbers every ninety days amid an unpredictable advertising market.

Zuora works with all sorts of SVOD services, from major cable networks to regional providers, and we're also lucky to work with the first dedicated online video subscription service ever. What's its name? Crunchyroll. More than a million subscribers pay Crunchyroll to watch hit anime shows like *Cowboy Bebop* and *Dragon Ball Z* (for my fellow uncool people, anime refers to Japanese animated TV shows), as well as exclusives like the anime prequel to *Blade Runner 2049*. Crunchyroll is actually big everywhere *except* Japan—they specialize in overseas rights and have viewers in more than 180 countries, from Brazil to Botswana.

Crunchyroll started off in 2006 as a piracy site. Not to condone piracy, but by the time they relaunched as an official subscription service in 2009, Crunchyroll benefited from two things: early brand recognition and a keen sense of what their fans liked and didn't like. They were the first dedicated online video subscription service, at a time when Netflix was still doing a lot of DVD business. And much like Netflix, today a significant percentage of the revenue Crunchyroll gets from its subscribers is plowed right back into investing in new content and supporting the anime industry in Japan.

Today the Crunchyroll "user conferences" look like giant Comic-Con conventions. Can you imagine a network television company filling a convention center with thousands of enthusiastic fans dressed as their favorite sports announcer from *Monday Night Football* or coroner from *Law and Order: SVU*? Me neither. Today they have more subscribers than all the people who bought the top five anime releases in history combined. "The more niche you are, the more you have to differentiate through some kind of aspect, and we do that through community. We have a very big brand team who focuses on engaging our audience," Crunchyroll marketing lead Reid DeRamus said to the editors of our *Subscribed Magazine*. "There's a lot of overlap in gaming, Twitch, e-sports, Comicon, a lot of overlap with that kind of crowd. We have a lot of hardcore anime fans on our site, who drove the growth of Crunchyroll initially, but we have a lot of titles that also appeal to the broader audience. It's not just die-hard anime fans anymore. It's like a beautiful blend of all of them." When you're trying to cut through, building community is key.

And what's missing among all these dozens of genre SVOD channels? Sports, of course. There's a huge race on right now to become the "Netflix of sports"—a single destination where fans can find live games of all their favorite teams and leagues, as well as commentary and highlight shows. Right now the leading contender is DAZN (pronounced "Da Zone"), a UK-based sports-only streaming site owned by the Perform Group. Founded in 2015, it's already live in Germany, Switzerland, Japan, and Canada, and hosts more than eight thousand sporting events a year for $20 a month, significantly less than the standard cable packages. "We've always had one eye on direct to consumer subscription play," CEO James Rushton told *Sportsmail*.

Much like Crunchyroll, DAZN is navigating the complexity of digital rights to bring compelling content to all sorts of new overseas markets. Lots of Canadians love the NFL. Lots of Japanese people love the NBA. Lots of Germans love English soccer (or football—forgive me, people of Britain). DAZN caters to those underserved markets. It helps that it's well funded—it recently bought domestic rights to Japan's top soccer league for close to $2 billion, and in Germany it won English and Champions League soccer rights away from some big cable companies. Think about that—a $20 a month streaming site is competing (and winning) for sports rights against major European cable networks like Sky Sports. For the first time in Europe, some Champions League games can be watched only on the internet! DAZN understands that in a connected world, there are huge opportunities up for grabs in undervalued international audiences. Let a thousand niche Netflixes bloom.

CORD CUTTERS ARE GREAT NEWS FOR THE CABLE INDUSTRY

Sports was supposed to be the glue that would hold the cable bundle together. That's clearly not the case anymore—live games are showing up on social networks, and lots of major league teams have their own SVOD services. The troubles at ESPN have been well documented—the shift to streaming services and subsequent loss of more than 13 million subscribers since 2011 came shortly after ESPN borrowed heavily to pay ridiculous amounts of money for television rights. As I write, ESPN is addressing these systemic challenges with layoffs, which is depressing, but a dedicated ESPN SVOD service is just around the corner. "Identifying the root cause of the acceleration in cord-cutting isn't hard. It's not demand (the demand has *always* been there). It is supply. Would-be cord-cutters and cord-nevers are finally being given options," analyst Craig Moffett told *Recode.*

According to Digital TV Research, SVOD revenues in Canada and the United States will reach $24 billion in 2021, up from $2.6 billion just five years ago. Roughly half of millennials and Gen Xers don't watch any traditional TV at all. These figures are causing lots of hand-wringing in media boardrooms, but I think that cord cutting may be the best thing that has ever happened to the cable industry. That might sound like a counterintuitive assertion right now (a Google search on "cable industry" prompts the autocomplete suggestion "dying"), but smart media companies stand to benefit enormously from the shift from coax to ethernet. Why? Once the shift to digital is complete, these businesses will be able to explore entirely new ways of taking advantage of their core assets (infrastructure, pipe, and people) in order to bring new services to their customer base. Stuff we haven't even thought of yet.

There are more than 19 million broadband-only homes in the United States right now, a number that is expected to nearly double by 2022 according to a recent report by Kagan, a media research company. Yes, the providers are taking a hit on cable subscription revenue, but their margins have always been better in broadband. Remember, those customers aren't going anywhere—they're just demanding different digital services. Saying goodbye to the bundle might be painful in the short term (most cable subscribers watch only 9 percent of what's available anyway), but it will eventually unlock more focused revenue streams. It's not like we've lost our appetite for video content! Cable companies still have a direct pipe into our living room, as well as huge infrastructures and employee bases (Comcast, Cox, and Time Warner employ more than two hundred thousand people). Smarter usage-based billing and cloud-based updates will make their video content services more responsive and valuable. They also have the opportunity to become the operating system of connected homes. In a few years we could be using our former "cable company" to upgrade an alarm service, schedule a new refrigerator installation, or discover we have some loose shingles on our roof.

Today, SVOD subscriptions generate more than $14 billion in video revenue compared with nothing ten years ago. Nearly half of American online shoppers pay for streaming media services. That's pretty amazing. As we transition through the efficiency stage of SVOD, we're starting to see glimpses of the opportunity stage. More media start-ups like Molotov in France are going to transform the way we watch television via cloud-based DVR services with powerful search tools and discovery algorithms. More former "movie stars" like Will Smith are going to start debuting new projects on SVOD libraries. And more former "movie producers" like Jeffrey Katzenberg are going to start launching subscription-based short-form video series with top-shelf production values, instead of lashing themselves to the masts of splashy sink-or-swim blockbusters.

STEVE JOBS VERSUS PRINCE

Today more than 30 million people in the United States pay for a music streaming service, and those services now represent more than half of the US music business. All that listening and discovery is having all sorts of positive ancillary effects—retail sales are up this year after fifteen years of decline. Ex–Sony Music CEO Edgar Berger sounded notably upbeat about the promise of paid streaming subscriptions, as well as the music business in general. He told *Billboard*: "The current trajectory is that the industry will inevitably grow; there's no doubt that paid subscription will be the predominant format in the market, the one that consumers will gravitate towards. The music industry is managing three transitions at the same time: from physical to digital, PC to mobile, and download to streaming. In that context I think the industry is performing remarkably well, and with a paid subscription model we are building a business that is here to stay."

Speaking of the decline of iTunes-style downloads, while Steve Jobs got most things right, he famously got it wrong about streaming services. "The subscription model of buying music is bankrupt," he told *Rolling Stone* in 2002. "I think you could make available the Second Coming in a subscription model and it might not be successful." That same year David Bowie made a much more prescient statement: "Music is going to become like running water, or electricity." Bowie was an early pioneer of connecting directly to fans through digital subscription services—he gave his fans exclusive tracks, photographs, videos, as well as Web space and email addresses through his own ISP service, BowieNet. Another artist who saw the changes coming? Prince.

Prince launched the NPG Music Club, an online music subscription service, on Valentine's Day 2001. In a way it was a precursor to Tidal. For five years, NPGMC (named after Prince's backing band, the New Power Generation) offered a monthly or annual membership that not only let fans get new releases, but also provided access to prime concert seats and passes for events like sound checks and after parties. We had Sam Jennings, his digital producer, on our *Subscribed* podcast, and he detailed just how committed Prince was to creating a sense of value around his service:

"They were getting about three or four new songs every month, live versions, remixes, all kinds of things. Plus an audio show. We called it an audio show but it was basically a podcast! It was essentially an hour-long radio program that Prince put together in his studio that we provided as a download. The idea was to create an ongoing experience for them, so that they want to be a part of it. They get the music, they get the downloads, but they're also investing in a larger experience, which is the community of subscribers themselves. The question was how do we make them feel more like members, and less like customers?"

And what if your listeners aren't just members, but participants in the creative process? In 2016 Kanye West dropped a new album, sort of. It wasn't actually finished—he kept publicly tweaking lyrics, rearranging the song order, and adding and subtracting material. As I'll explain in greater detail later in the book, in the technology industry we would call *The Life of Pablo* a minimum viable product. That may sound like a pejorative term, but a minimum viable product is actually incredibly important. Only after it gets something out in the market can a business gather customer feedback and use this data to iterate and improve in a continuous deployment cycle. The MVP is a defining principle of cloud software development, and Kanye applied it to his music-writing process.

What happens when a static product like an album turns into a fluid service like a music stream? All sorts of interesting things. Today thousands of musicians are benefiting from platforms like Patreon that give them a steady, dependable source of recurring revenue. Much like Eric Ries's "Lean Startup" method, they're shortening their product development cycle through experimentation, validated learning, and iteration. And they're creating a virtuous feedback loop whereby customer responses help inform product development. By putting it out there—and letting subscribers pay for it—these musicians are successfully feeding their sales funnels without having to wait for a finished product (although I'm sure they wouldn't put it that way!). Instead they can tinker with their music, optimizing as part of an ongoing deployment cycle.

Finally, if there was any musician who exemplified the virtues of constant iteration and experimentation, it was Prince. After he shut down the NPG Music Club, he sent out the following email to his fans. I think it does a great job of capturing his artistic genius, relentless curiosity, and willingness to cast off the past to begin again. Overall, it's a wonderful testament to creative freedom.

Greetings Family,

The NPG Music Club has been in existence 4 more than 5 years. In that time we've learned a great deal from each other and about this brave new online world we have all chosen 2 b part of. The members we have been 4tunate enough 2 have join r family have truly made this the best music club any artist could ever dream of. And all the things we have been a part of 2gether—the concerts, the celebrations, the soundchecks, the discussions and the un4gettable music—have shown us what a New Power Generation can truly b. We thank u from r hearts 4 sharing urselves and ur love of the music with all of us. It has been a blessing.

Once the NPG Music Club won the 2006 Webby Award, discussions within the NPG began 2 center on what was next. What's the next step in this ever-changing xperiment? The achievements of the past cannot be questioned and we r truly grateful 4 everything that has been accomplished. But in its current 4m there is a feeling that the NPGMC gone as far as it can go. In a world without limitations and infinite possibilities, has the time come 2 once again make a leap of faith and begin anew? These r ?s we in the NPG need 2 answer. In doing so, we have decided 2 put the club on hiatus until further notice.

The NPG Music Club was a first step; the lessons learned will last 4ever. Now comes a time of great reflection and re-structuring. The future holds nothing but endless opportunity and we plan on seizing it wholeheartedly. Don't u want 2 come?

Love4oneanother, NPG Music Club 4ever

CHAPTER 4

PLANES, TRAINS, AND AUTOMOBILES

You don't buy Hyundai's new hybrid car the Ioniq—you subscribe to it, for $275 a month. It's a lot like picking a cell phone plan: pick your model online, choose between a twenty-four- or thirty-six-month plan, select your upgrades, then walk into a dealership to pick up your vehicle. No price haggling, no loans, no back-office pitches. "Our goal is to make car ownership as easy as it is to own a mobile device," says Mike O'Brien, vice president of product planning for Hyundai. "Instead of all the steps you go through to purchase—you've got to find financing, you've got to negotiate the deal, you've got to think about the trade-in—all those steps are very complicated, particularly when you talk to millennials." Personally I would replace the phrase "very complicated" with "purposefully tortuous."

Hyundai has lots of company when it comes to automobile subscriptions. Porsche's Passport subscription features access to half a dozen car models and covers maintenance, insurance, and vehicle tax and registration starting at $2,000 a month. Cadillac offers access to its current vehicle lineup for $1,800 a month and lets you switch out vehicles as frequently as eighteen times a year. Subscribers to Ford's Canvas program pick out a monthly mileage plan and can roll over unused miles into the next month, much like a data plan. You can subscribe to a Volvo XC40 (their compact SUV) for $600 a month, and that includes concierge services like packages delivered straight to your vehicle. Everything is covered except the gas: insurance, maintenance, wear-and-tear replacements, 24/7 customer care. Volvo's CEO expects that one out of every five of the company's vehicles will be delivered via subscription by 2023, and the company is working on its own ridesharing network that will allow users to loan or rent its cars for profit. Jim Nichols, product and technology communications manager at Volvo USA, told *Consumer Reports*, "Our research has shown that many customers are looking for a hassle-free, fixed-rate experience that mirrors the many subscriptions they currently have, such as Netflix or Apple's iPhone [upgrade] program."

But wait—isn't a vehicle subscription just another word for a lease? Well, no. A lease still binds you to a specific vehicle, whereas a subscription can potentially offer you access to a range of vehicles. "Simply flip between vehicles via the app as your needs change," says Porsche on its website. You're signing up with the company, not the car. Another difference: With subscriptions, all the potentially annoying aspects of owning a vehicle (registration, insurance, maintenance) simply go away. With leases, you still have to get your own insurance. Also, many car subscriptions give you the option to subscribe on a month-to-month basis. As Christina Bonnington of *Slate* notes, "You could theoretically not have a car for ten months of the year when you're working and using public transit and then get a car subscription for two months when you'll be travelling more often." Subscriptions also don't offer the option to buy when they conclude, which I view as a huge positive—that means that it's in the carmaker's interest, not yours, to keep its vehicles in great shape.

It's not just millennials who think that car ownership is expensive and onerous. When I was just out of college, my used car was a ticking pecuniary time bomb that threatened to wipe out my savings account as soon as it broke down. The US auto loan market is currently a gigantic trillion-dollar beast. Here's my prediction: The vast majority of that money is going to pour into subscriptions and car services over the next ten years. Car companies are clearly responding to a sea change in consumer preferences toward services, but they're also responding to another massively disruptive phenomenon. What would that be? Uber, of course.

THE BIRTH OF RIDESHARING AND THE END OF TRANSPORTATION AS WE KNOW IT

Let me back up. During the first couple of years of Zuora, when we were trying to convince other people that subscription models weren't just for software companies, we were fascinated with a company called Zipcar. Founded in 2000, Zipcar let subscribers book cars by the minute, hour, or day. It pitched itself as an alternative to car rentals or U-Haul. It was a novel service that was also simple and intuitive. Zipcar had several thousand cars scattered throughout twenty-five major American cities. You located a nearby Zipcar and reserved it online, swiped your membership card across a sensor on the vehicle, and drove off. It was also very popular—by 2012 it had more than three quarters of a million drivers paying for transportation by the hour. At one of our earlier events in New York City, for example, we found out that no one had a car—no surprise there for anyone who's lived in New York. But what came as a surprise was that 80 percent of people we polled had Zipcar memberships. Yes, there were massive limitations to Zipcar—you had to live in a city, for example. But we could see that the next revisions of this concept (give me the ride, not the car) were just going to get better and better. That experience let us see a future world where car ownership would not be necessary.

Today more than 60 million riders use Uber and Lyft. These ridesharing services have ushered in a whole new set of consumer priorities: Why buy a car at all, when all you need to do to get from point A to point B is pull out your phone? Why can't I just subscribe to transportation the same way I subscribe to electricity and internet access? But wait, you might say. Uber isn't a subscription service—there are no monthly fees. I disagree. It sure looks and feels like a digital subscription service to me. Uber has your ID and all your payment particulars, and it employs usage-based pricing so that you pay for only what you use. It knows your usage history (your home, your work, your common destinations) and uses that information to customize its service for you. And thanks to its partnership with Spotify, it even knows your favorite music.

Oh, and guess what? Uber does in fact offer monthly subscriptions. Right now Uber is testing a flat-rate subscription service in several cities. Users can pay a monthly fee in exchange for bundles of reduced-rate trips with no surge pricing. In other words, Uber will cut you a deal on rides in exchange for steady business. The company may take a short-term profitability hit, but the goal is to gain long-term customer loyalty in a very young and turbulent market—and this customer loyalty is becoming more and more important as ridesharing becomes a commodity. Here in the Bay Area, the Uber and Lyft markets are really fluid. I'll frequently toggle between the two services—lots of the cars even feature both logos in their windshields. There's very little brand loyalty on my part.

Now contrast that with my Amazon Prime experience. All due respect to other potential ecommerce vendors, but Amazon has my business, in no small part due to Amazon Prime—they hooked me with the free shipping, and now I've got music, movies, and all sorts of other services. I'm not going anywhere. Uber and Lyft are both vying for that same lock-in effect by offering discounted services around consistent consumption patterns—in other words, they're going after my commute. As Lyft president John Zimmer, anticipating fully autonomous vehicles, told *The New York Times*: "The cost of owning a car is $9,000 a year. Let's say we offer a $500 monthly plan in which you can tap a button and get access to transportation whenever you want it, and you get to choose your room-on-wheels experience. Maybe you want a cup of coffee on your way to work, or you want to watch the Warriors game later, so you're in what's basically a sports bar, with a bartender."

CELL PHONES ON WHEELS

Today the accepted Silicon Valley wisdom is that as cars turn into cell phones on wheels, software will inevitably trump hardware, just as Microsoft trumped IBM. As lithium batteries replace combustion engines, automobile hardware will become commodified, and the new growth market will be in information services. The analyst firm Gartner predicts that there will be 250 million connected cars on the road by 2020. That means one in every three cars on the road will be connected. By then, digital diagnostics, infotainment channels, and enhanced navigation systems are expected to constitute a $270 billion industry, up from $47 billion today. At some point, the data and services associated with a vehicle may be worth more than the vehicle itself (much like a cell phone!). GM's OnStar, for example, which started in 1996 as a concierge service, is now in more than 12 million vehicles and hosted more than 1.5 billion customer interactions last year.

A lot of Silicon Valley executives think that the legacy car companies today look a lot like IBM did in 1985. That year, Big Blue employed more than 400,000 people (more than three times the size of Apple today). Digital Equipment Corporation, its nearest rival, had a quarter of its employee base. Apple itself was in a tailspin, having just fired Steve Jobs in one of the most infamous board decisions in history. Personal computers were an admittedly low-volume market at the time, but the IBM PC dominated it. Compatibility with its PCs was an industry mandate. It had clear market superiority. IBM had the hardware wrapped up; it just needed a decent graphical user interface.

The same year, Microsoft, which had roughly two thousand employees, introduced Windows. That was the beginning of the end for the IBM PC business. By the early 1990s IBM-manufactured PCs were the exception rather than the rule, and the industry revolved around Windows. By the end of 1996 Microsoft had a market cap of $98 billion, while IBM was worth $80 billion. As analyst Horace Dediu notes, "Without control over the platform, PC hardware is nothing more than a commodity, with negligible margins, intense competition and an inability to control one's destiny." IBM lost the war by giving the user experience over to Microsoft, and the same thing is going to happen to the legacy car companies when they inevitably hand over their dashboard intelligence to Apple, Google, and Facebook. And with all the dramatic improvements we're seeing in manufacturing and 3D printing, maybe a whole new crop of automobile start-ups will be able to batch-print their own vehicles in Chinese factories (just like cell phones!), right?

Wrong. As it turns out, *it is really hard* to build a safe, great car at scale. Just ask Elon Musk. Or Apple. Or Google.

SILICON VALLEY VERSUS DETROIT?
BET ON DETROIT

As it turns out, the Big Three have some distinct institutional advantages over Silicon Valley when it comes to building the future of the automobile industry. First, they have the distribution. The vast dealer networks these companies operate are commanding assets. Second, the scale of their operations is impossible to duplicate; more than 17 million cars were sold in the United States last year (Tesla sold around 100,000). Sourcing and assembling vehicles involves extensive regulations, and the margins aren't great. Companies need to invest billions of dollars in factories and distribution channels to make the whole production process work.

Also, the three largest automakers' financial resources are huge. Since GM and Chrysler emerged from bankruptcy in 2009, the Big Three have invested more than $30 billion in new jobs and facilities. The American automobile industry spends $18 billion a year on research and development, focusing on fuel-efficient, electric, and autonomous vehicles. GM CEO Mary Barra says that her company is "quarters, not years" away from deploying fully autonomous vehicles at scale. These car companies have spent decades crafting their vehicles and their brands, and as a result enjoy some forbidding advantages, but there is a red flag—if they don't know their drivers by the time autonomy and access-based consumption roll around, they will lose out to a competitor.

Finally, the Big Three are currently in the midst of reimagining themselves as not just car manufacturers, but transportation solutions. They get that automation is coming. They get that in the future they'll probably be doing more fleet management and fewer individual vehicle sales. They also understand that they'll still be the ones making the vehicles for all these new ridesharing ser-vices. And they also recognize that true "mobility as a service" en-tails taking advantage of all sorts of modes of transport, not just driving. And they see that as a huge opportunity.

Daily transportation is still drowning in pain points. Most of us are just hoping to get from our bed to our desk and back again with the least amount of hassle. As Jamie Allison, director of connectiv-ity and mobility for Ford, said at our *Subscribed* San Francisco con-ference, his company's new mandate is to make that "bed to bed" journey as simple as possible, which explains Ford's investments in Chariot commuter vans as well as the massive expansion of its bike-share program. So much of getting from A to B involves negotiating a mind-numbing series of transactions: fines, tolls, leases, tickets, repairs. What if you had one ID to handle all of those logistics?

"We are seeing an evolution toward services rather than physi-cal transactions," Allison said at our conference. "There's been a fragmentation of the customer experience. When you bought a car, or you leased a car, that was one interaction. Then as you went and needed service for your car, there'd be a different interaction at the dealership. And for all these years, the cars weren't con-nected, so we really didn't have a view of the journey that you're on." Allison is absolutely right about the fragmentation of the cus-tomer experience—most of the big auto manufacturers conceded the service aspect of their industry to thousands of dealerships and repair shops a long time ago. Today Ford is actively trying to remedy that. As Allison noted, "FordPass is a portal to a seamless customer experience." FordPass app users can warm up their cars in the driveway on cold mornings, find and reserve parking spots, schedule service appointments, find nearby gas stations, and make mobile payments.

Henry Ford had a famous quote: "If I had asked people what

they wanted, they would have said a faster horse." Today Ford understands that it can't solve for mobility just by selling more cars. Its focus on urban dwellers is smart—today more than half of the world's population lives in cities, and two-thirds will by 2050. Right now Ford has around 6 percent of the $2.3 trillion global automotive market, but close to nothing in terms of the $5.4 trillion transportation services market. It also understands that those two markets aren't mutually exclusive, and there's a huge amount of opportunity ahead, as long as it starts first and foremost with the needs of the passenger, then works backward from there.

ALL YOU CAN FLY

You may be less aware of this, but similar changes are happening in the airline industry. We all know what flying is like: awful. Even for frequent business fliers like me, getting into a plane seat is never a pleasant experience. Enter Surf Air. Former president and CEO Jeff Potter previously managed an airline and a vacation destination club, and he decided to combine both concepts. The result is Surf Air, which is often called the "Netflix of Aviation" or the "Uber of the Skies." Its members get access to limitless flights for a flat monthly fee—right now they're in the western United States and Europe, and growing rapidly.

This is another classic example of building a successful business by starting with the customer's wants and needs, attacking pain points with a machete in order to surface the best outcome, and growing a loyal subscriber base in the process. With Surf Air, you take out your phone, find out when the next few flights are leaving, and reserve a seat. As a member you're prescreened, so you just walk onto the flight. For the people in my company who travel around the West Coast all the time, it's been transformative.

The airline industry has been ripe for disruption for decades, with customers demanding greater autonomy and control over their experiences. Air travelers are constantly being fleeced when they try to buy last-minute tickets—why should travel planning flexibility be penalized? And think about all the time we spend hanging around in airports—the inefficiencies of plane travel might be benefiting the airport retail industry, but most of us would just as soon spend that time somewhere else. There are more than 200 million frequent fliers worldwide, and as customer preferences shift, every one of them is up for grabs.

Eventually, the current airline industry strategy of shameless fee charging is going to collapse under its own weight. It's the depressing result of a product mindset that prioritizes add-ons and revenue extraction and devalues customers. What could a flying experience look like in the future? Well, to start with, it might also include cars and trains. Maybe United sends you a cobranded Uber car with a monitor that includes all your hotel and flight details, a drop-down menu to preselect all your entertainment and dining options, and light rail information for your destination city. Maybe that car's arrival time at your house is synchronized to your flight's actual departure time. Maybe you can start binge-watching *Narcos* in the car and pick it up on the plane where you left off. Maybe when you arrive at the airport, a service like Clear can speed you through security lines with a swipe of your boarding pass and a thumb scan, because all your standard ID information has already been paired with your biometric details. Maybe all these services could be wrapped up in a flat annual frequent-flier membership plan.

"From a business perspective, we all know that airlines have struggled for years," says Mac Kern, former vice president of commercial planning at Surf Air. "It's a very capital intensive business, not to mention commodity-based. Prices get driven downward. It's very competitive. The subscription model gives us predictive revenue—that's something that no commercial carriers have. They don't know if a flight is going to be profitable until the door on the airplane closes (and they still have to fly at that point!). Because of subscriptions, we know exactly how much revenue we're going to generate at the beginning of every month. So we can scale our operation effectively, because we know exactly how much flying we're able to execute. That kind of insight is basically magic in the aviation industry. No one has been able to do that before."

Today there are airline companies, telecoms, streaming music services, and newspaper publishers all asking the same kinds of questions: What's the value of this new service (or route) to our subscriber base? Is it receiving the kind of support that we predicted it would? How long are our members staying with us? What does our growth efficiency look like? What do our usage patterns tell us about where to apply more resources? Who might be at risk of churning? Right now Surf Air costs around $2,000 a month, so it's specifically aimed at frequent fliers with more money (or access to expense accounts), but we're going to start seeing this model spread through the entire aviation industry in much the same way it's transforming the automotive industry. It's just a matter of time.

RAILWAYS AND RIDESHARES

Competition in the transportation industry has switched from vertical to horizontal. By that I mean it's not just car companies going after car companies, or airlines taking on other airlines anymore. Light rails are competing against rideshares that are competing against budget airlines. Everyone is chasing after passengers who are demanding access to anytime, anywhere travel. Take SNCF, France's state-owned railway company, which was founded in 1938. If you've ever gone backpacking in Europe, you've ridden SNCF. And for lots of young French people working in cities, SNCF has been a reliable way to visit home on the weekends. But over the last couple of years, they've seen some fierce competition from new ridesharing services, long-route coaches, and discount airlines.

SNCF realized that it needed to compete against all these new transportation platforms to stay viable, so it decided to launch a new targeted subscription service: young adults between the ages of sixteen and twenty-seven would get to ride anywhere they wanted, for 79 euros a month. Once it got the idea, it took eight months for SNCF to roll out the program. That's ridiculously fast. Signing up for its old card-based loyalty program used to take three weeks for passengers to complete. This one took five minutes. The results were staggering. Today 75,000 more young French people are taking the train. SNCF hit its annual growth targets within a couple of months.

"The model of unlimited subscription has become a reality in telecommunications, sports halls, and cinema," says Rachel Picard, general manager of Voyages SNCF. "Young people want more freedom, they like to travel, but they decide on things late and tend not to book in advance for cheaper rates. So the idea was: Why don't we use online subscriptions in order to adapt to these new modes of consumption and travel habits? Why not give them unlimited rides?" Why not? Based on behavioral data from existing monthly membership card programs, SNCF had a good idea to spin up a new subscription initiative. And with the right back-end systems, it was able to launch this initiative quickly.

Think about it. This is an eighty-year-old state-owned rail company competing against transportation start-ups like BlaBlaCar, a wildly successful French ridesharing platform that connects passengers with empty seats on long-distance car journeys, allowing passengers and drivers to share the cost of a trip. It did this by taking advantage of a cloud-based suite of solutions: Salesforce for CRM, Zuora for subscription management, SlimPay for payments, and AriadNEXT for document verification, among others. SNCF is yet another example of a big "incumbent" company reacting nimbly with the help of a few new software services. And since it was targeting a young audience, it also put lots of resources into social media monitoring and marketing. SNCF knew from the start that its customer service platforms were going to be Twitter and Facebook. The result was thousands of young people seeing their families and loved ones more often.

Trains, bike shares, subways, shuttles, and car services are all locked in horizontal competition, but smart partnerships and platforms will help commuters carry their identity across all these networks seamlessly and intuitively. The winners will be the services that don't just manage routes, but solve for A to B. Helsinki is experimenting with a mobility planning app called Whim that generates instant itineraries that mix private and public transportation networks, and can even recommend healthier routes when the weather is nice. As *The Economist* notes, "Young urbanites, who have become accustomed to usership instead of ownership, find the notion of transport as a service both natural and appealing. Meanwhile the cost of running a car in a city goes ever upwards. Parking gets harder. Many city-dwellers are questioning whether the convenience is worth it. Between 1983 and 2014 the share of Americans aged 20–24 with a driving licence fell from 92% to 77%."

If there's anything that these stories can tell us, it's that transportation is rapidly evolving from a sequence of painful but necessary transactions toward an intuitive service that will eventually become seamlessly embedded in our daily lives. Even the whole idea of building specific industries around specific vehicles is being called into question. Car companies are buying bike and shuttle companies—who's to say that an airline won't be next? Meanwhile, airline executives are swapping notes with digital media experts around topics like churn, retention, and customer lifetime value. Train companies are moving beyond thinking about customer segmentation and into thinking about individual customers themselves. Things are becoming unstuck. It's pretty exciting.

CHAPTER 5

———

COMPANIES FORMERLY KNOWN
AS NEWSPAPERS

So much for the death of the newspaper industry. A recent Nielsen Scarborough study found that more than 169 million US adults now read newspapers every month, in print, online, or on a mobile device. That's almost 70 percent of the adult population. According to the latest Reuters Institute Digital News Report, *The New York Times*, *The Wall Street Journal*, and *The New Yorker* all gained hundreds of thousands of new digital subscribers in 2017. *Vanity Fair* picked up 13,000 new subscriptions in *one day*.

Remember when millennials would never pay for content? According to the Reuters Institute, in the United States, the proportion of people ages eighteen to twenty-four paying for online news leaped from 4 percent in 2016 to 18 percent in 2017. *The New Yorker*, for example, has seen its new millennial subscribers more than double from the same period a year earlier, and *The Atlantic* has seen its new subscribers age eighteen to forty jump 130 percent. Even free news outlets are seeing a bump in paid contributions—

The Guardian's charter stipulates that it provide free news, so it has been experimenting with voluntary donations in the form of memberships, and it's been a huge success. As of March 2017, *The Guardian* had sold more than 230,000 memberships at between £6 and £60 a month, along with getting 160,000 one-off donations. According to Reuters, many more Americans are paying for online news than ever before—roughly 16 percent of the population, a 7 percent jump from 2016 to 2017.

Sure, those papers can thank a certain controversial head of state for much of their new business, but this isn't just a political story. Digital subscriptions are transforming the broader publishing industry in profound ways, and a new breed of reader-supported titles are enjoying newfound popularity. In the technology industry, for example, Jessica Lessin's sharp, pointed (and subscription-only) *The Information* now has the second-largest team of tech reporters in Silicon Valley. Ben Thompson has thousands of readers who are happy to pay him $100 a year for his excellent *Stratechery* newsletter, and Bill Bishop writes an email newsletter about current affairs in China called *Sinocism* that has more than thirty thousand readers paying $118 a year. Meanwhile, all sorts of splashy, venture-funded, "digitally native" titles like *BuzzFeed, Mashable, The Daily Beast,* and *Vice* are struggling to hit their numbers. Any guesses why?

THE DECLINE AND FALL OF AD-SUPPORTED JOURNALISM

Remember when the prevailing consensus was that the newspaper industry was in a terminal death spiral? "Newspapers are dying," intoned *The New Yorker* in 2008. "The evidence of diminishment in

economic vitality, editorial quality, depth, personnel, and the overall number of papers is everywhere. What this portends for the
future is complicated." Craigslist, the Great Recession, and a plummeting print advertising market were all conspiring to massacre
the Fourth Estate. At the same time, the venture capitalists funding all the cool new ad-based news sites like *BuzzFeed* and *Vice*
were very happy to lecture the newspaper industry about the new
way of conducting journalism—goodbye to local monopolies running on classified ads, hello to zero distribution costs and free content. Unburdened by printing presses and delivery trucks, the new
digital news sites were going to "go broad" and sweep up all those
exploding digital advertising dollars (wait a minute, wasn't this the
same kind of thinking that caused newspapers to launch free sites
twenty years ago?).

Why are readers and publishers alike embracing paid subscriptions for content services over ad-based business models? There
are several reasons, but the dismal state of advertising is a big one.
Of the thousands of potential reasons why advertising is terrible,
let me offer you three. First, people don't like it. According to Me-
diaPost, roughly a quarter of all Americans use some kind of ad
blocking software, costing publishers almost $16 billion in annual
revenue. Around a quarter of the Reuters Institute's survey of seventy thousand readers around the world use ad blockers on a daily
basis (Greece has the highest penetration at 36 percent and Korea
is the lowest at 12 percent). And despite a lot of talk about "smart
targeting" and "native advertising," we're all still drowning in
thousands of commercial messages a day that lack any kind of rel-
evance to ourselves or our daily lives.

Second, digital ads also don't make much sense from a business
perspective. According to the Interactive Advertising Bureau, of
all the online advertising dollars spent in 2016, 49 percent went to

Google, 40 percent went to Facebook, and 11 percent went to "everyone else." As Josh Marshall of *Talking Points Memo* describes it: "There are too many [digital] publications relative to advertising revenue. So let's imagine there are 30 publications and 25 revenue seats. The publications fight like hell to secure one of the seats. Then the platform monopolies came along and sat down in maybe 5 or 10 of the 25 seats. You can see the problem. The competition of 30 publications competing for 15 seats gets insane. A bunch of the publications are going to die or be forced to find another way to fund themselves." And, of course, this entire *Hunger Games* scenario can also quickly implode once the advertising market turns south after a recession.

Finally, ads have all sorts of other insidious effects, like turning content providers into clickbait factories. Ads fund the race to the bottom. Ex–*Politico* president Jim VandeHei calls it the "crap trap." As Jessica Lessin of *The Information* says: "I still believe it's much safer to build a business that doesn't need any advertising to survive. Doing so forces you to focus 100% on your value to your readers. It's the only way to make sure that what the news publishers deliver to readers in the future is smarter, more informed and more relevant than in the past." In fact, I'd argue that back when newspapers were drowning in classifieds and advertising income, their business models were very much driven by a myopic product mindset, on both the print and digital side, that took a real toll on readers. As always, the goal was to shift as many units as possible, from print copies to page views, to boost those ad inventory rates. Sifting through a Sunday print paper that featured twelve pounds of coupons and brochures used to be a pretty grim exercise, not to mention trying to navigate through a "free" newspaper website that was plastered with pop-ups and slide shows.

Given that ad revenue is notoriously inconsistent, what else is going on? At the same time that publishers are giving the broken ad system a hard look, there's a whole new generation of consumers who are comfortable subscribing for services—Spotify, Netflix, food boxes, productivity apps—as long as they stay timely, relevant, and focused. It's not surprising that the Nordic countries, where Spotify and OTT content services have a much stronger foothold, have the highest percentages of people paying for online news (15 percent in Norway, 12 percent in Sweden, 10 percent in Denmark, 7 percent in Finland). Other platforms like Patreon are helping people directly subscribe to hundreds of new video shows and podcasts with regular monthly payments. There is a strong correlation between people who are willing to pay for streaming services and people who pay for online content: "Other online services have basically given people the grammar with which to understand subscriptions," says Nic Newman, lead author of the 2017 Reuters Institute Digital News Report. This is a marked shift for an industry that once relied on advertising for well over half of its revenue mix.

Advertising is never going to go away, but as digital subscription services become the norm, readers and publishers alike are starting to appreciate the dividends of direct reader relationships. Remember when paywalls used to be a controversial idea? Today the behavioral insights that come with membership plans and smart paywalls are helping publishers move away from empty-calorie metrics like CPMs and slide-show clicks toward more valuable engagement metrics like the amount of time spent on a site. Subscriptions make advertisements more relevant, and therefore more valuable.

"Making advertising a secondary—though still vital—revenue source is the most important strategic goal for most news publishers," says Ken Doctor, a very sharp critic of the industry who runs a great website called *Newsonomics*. "Reader revenue, if backed by sufficient high-quality content and good digital products, proves far more stable than advertising." Of course the newspaper industry still faces headwinds as it shifts from a print ad model to one largely driven by digital subscriptions, but today's consumers are increasingly comfortable with supporting smart digital services of all kinds. According to the Reuters Institute report, Americans who already pay for streaming media services are five times more likely to pay for online news than people who don't pay for online media.

THE MYTH OF PRINT VERSUS DIGITAL

Today's successful publications are shifting from thinking about how to monetize the format—the feature, the banner ad, the slide show—to prioritizing the wants and needs of the reader. You can see how this "format blindness" played out during the debate over micropayments a few years ago. Newspapers were going to charge people a nickel or a dime per article, which never made sense to me—that put a premium on individual pieces of content versus the overall brand (you can see this same debate being played out today with Facebook Instant Articles). As it turns out, the papers had lots of loyal readers right under their nose—their home delivery customers. Look at the way Londoners are loyal to their papers. On a recent flight I overheard a Brit ask for a copy of *The Guardian*. When the flight attendant apologized and offered him a copy of *The Daily Telegraph* instead, his response was priceless: "Why should I read that fishwrap?" The papers realized that loyal readers were willing to pay for smart writing; they were simply doing it on their phones instead of wrestling with newsprint on the trains, so digital subscriptions suddenly became really important. Why one format should cost money while the other one shouldn't made absolutely no sense.

Print versus digital was always a false choice. The prevailing wisdom used to be that digital-native media companies like *Vox* and *BuzzFeed* operated in some magical overhead-free universe, while the anchor of sinking print advertising was dragging the newspapers down permanently to the bottom of the ocean. But the whole "print versus digital" argument assumed that the physical delivery of the content is more important than the content itself. The core of *The Wall Street Journal* is not its physical newspaper, but rather its journalists, its brand, its culture, its reach, its values. Its real value (and requisite expense) is in its content, not its format. That is something that people will pay for.

If you start with the premise that you have loyal customers who identify with your brand, and you see this as a chance to engage them more deeply with a rich digital experience, then your new model becomes less reliant on the whims of advertising and more supported by stable, recurring subscription revenue. In fact, lots of smart publishing groups are leveraging the loyalty of their existing print subscribers in their transition toward primarily digital revenue models. And they're taking advantage of the flexible pricing and packaging models and bundled add-ons to do it.

All of these publishers are rapidly turning into what Ken Doctor calls "companies formerly known as newspapers." These are savvy media organizations that are matching their core intellectual assets with complementary services to create new experiences for their readers, from small perks like complimentary Spotify accounts (News UK) and free business book downloads (WSJ+) to tee times, cruises, and conferences. In some cases, they're discovering new digital revenue models that are actually overtaking their print base.

THE ENTHUSIAST NETWORK: FUELING THE DRIVER'S JOURNEY

Take *Motor Trend* magazine, whose first issue appeared in May 1949. Today it's published by a company called The Enthusiast Network (TEN). Several years ago, the magazine started posting its website videos to YouTube. After more than a billion views and more than five billion free subscribers, in 2016 the company launched Motor Trend on Demand, a subscription streaming platform available for $5.99 per month that included hundreds of hours of exclusive content, including shows like *Roadkill Garage* and *Head 2 Head*. Today less than half of *Motor Trend*'s revenues come from its print edition, and it expects recurring revenue from its SVOD service to account for 20 percent of its overall revenue mix in short order.

That's a pretty astonishing transformation for a company "formerly known as an automotive magazine." In addition to syndicating its original content to places like Roku, Apple TV, and Amazon, it continues to experiment with different pricing models and subscription plans, wrapping in merchandise and passes to *Motor Trend* events—even throwing in free subscriptions to the print magazine! "The logic is inescapable. It's the way the modern media landscape is going," said Angus MacKenzie, chief content officer of TEN, to *Digiday*. "There are fewer and fewer instances where you are only in an ad-supported business model. A subscription platform gives us a consumer-supported business model—it's roughly analogous to what the magazines used to be."

Like all successful subscription services, TEN is taking advantage of predictable recurring revenue to stay razor-focused on its audience, create space to come up with distinctive new features based on popular topics (*The New York Times* now has a sizable revenue stream just from its crossword app, and it just launched a new paid cooking app), and avoid the commodification crap trap. We had a publisher of a magazine for equestrian enthusiasts at one of our *Subscribed* conferences in London who made a very salient point—surely there were other ways outside of a £20 a year magazine subscription to do business with a readership *that owns horses* (they got more engaged in competitions and accessories). Some of these initiatives succeed, some fail, but the goal of all of them is to drive legitimate reader engagement, as opposed to reverse engineering an editorial product for the benefit of advertisers.

BRITISH WITS: *FINANCIAL TIMES* AND *THE ECONOMIST*

Pricing agility—being able to pivot on a dime when a big story breaks—is also key to this transformation. *Financial Times* knew they were going to get a surge of traffic over the weekend of the Brexit referendum, so what did they do? Dropped their paywall for all their Brexit news and made sure that the flood of new readers saw plenty of tailored subscription offers. As a result, they saw a 600 percent surge in digital subscriptions sales compared with the average weekend. Today *FT* has more than 900,000 subscribers, with over 75 percent of their revenue coming from digital subscriptions.

Jon Slade, chief commercial officer of *Financial Times*, told *Digiday*, "We dialed up our marketing on a real-time basis. We were looking at buying patterns, opportunities in social, and spending our marketing budgets in pretty aggressive ways in an attempt to try and dominate a story. We then made sure that didn't conflict with the efforts of our audience engagement team, so there was constant dialogue between audience engagement and editorial, and between marketing and acquisition." There is at least as much innovation and creativity happening in *FT*'s acquisition efforts as there is in its exceptional journalism. *FT* also has a simple but brilliant formula for gauging reader engagement. Borrowing from the retail sector, they score every one of their readers on the multiple of three factors: recency (when did they last visit?), frequency (how often do they visit?), and volume (how many articles have they read?). Low scores indicate churn risks that their promotions group can approach with discount offers.

The Economist has had similar success with creative pricing strategies to drive acquisition. A few years ago, it made an impressive bet. Instead of going with the common practice of throwing in free digital content with a print subscription, it decided to charge for digital as well. And why not? Most large publications put a lot of work into their stand-alone digital efforts. That should be worth something, right? "The idea was to increase our renewal subscription revenue by making people opt for the print and digital package rather than the print-only package. So, instead of giving it away for free, we are actually starting to charge a premium for it," explained Subrata Mukherjee, former vice president of product and head of business systems at The Economist Group. It was a brilliant move, and it increased revenue 25 percent on those willing to add digital to print. *The Economist* is now experimenting with various ways to bring more people into the funnel, like presenting subscription offers to those who have ad blockers installed, special student packages, and frequent flier incentives.

Now, I understand that the newspaper industry certainly doesn't need any lecturing on the merits of the subscription model—they invented it, after all. And, of course, I recognize that the industry has been through an immensely painful period. I'd actually like to see more newspaper industry executives talking to SaaS executives about how to build and engage digital audiences. Newspapers have a lot to teach start-ups. A few years ago I saw a joke on Twitter from a media executive named Mark Lotto: "This is your semi-annual reminder that if *The New York Times* had been founded 5 years ago, not 160, it would be valued at $40 billion." There's some real merit to that observation—in fact, I'd argue that from a business model perspective *The New York Times* is starting to look much more like a SaaS company than a newspaper.

THE NEW YORK TIMES IS A UNICORN

You've probably heard of unicorns—not the mythical equestrian variety, but the Silicon Valley software companies valued at more than a billion dollars, thanks to record-setting investment rounds fueled by oceans of available cash and free borrowing. Of course, a few have left us, owing to "down rounds" (where companies raise at a lower valuation than their previous round), discounted acquisitions, or bumpy IPOs. Many if not most of these companies run on some form of subscription revenue. If *The New York Times* were a SaaS company instead of a newspaper company, I have no doubt that it would be worth twice as much as its current $4 billion valuation.

As of this writing, *The New York Times*'s stock price is at a five-year high, owing to the fact that today more than 60 percent of its revenue comes directly from its readers. Traditionally, newspapers have run on a greater proportion of advertising to subscription dollars, so *The New York Times* has reached a notable "crossover" moment with its revenue mix. That mix forms the basis of a strong, sustainable business model that will help it achieve its goal to generate $800 million in digital revenue by 2020 (it's already at $600 million in digital revenue, and overall subscription revenue is more than a billion dollars). And in the second quarter of 2017, for the first time ever, digital-only subscription revenue exceeded print advertising revenue. Expect that ratio to widen. And including its cooking and crossword apps, today the paper has more than 2.6 million paying digital subscribers. Ask any VC firm what kind of valuation they'd put on a SaaS business with those kinds of numbers!

The New York Times has successfully managed to flip the subscription/ad revenue ratio that most publications have on its head, which is a good thing—it protects the *Times* from the whims of the economy (and subsequently the advertising market), while also allowing it to charge a premium for access to a dedicated, professional audience. Roughly 4 percent of its total digital audience pays for content. I see that as an upside potential of 96 percent, but it's also a lesson that lots of SaaS companies here in the Valley have known for a long time—most of your profits are going to come from a minority of core users.

The closest analogy in the tech industry here is the "freemium" model, in which you offer a basic version of your service for free in order to acquire lots of users very quickly, then try to incentivize them with an enhanced paid plan. While the number of paying users is always going to be a fraction of your freemium base, if you have a great service, that core group can support your business—just ask Dropbox or Slack. As it turns out, loyal newspaper subscribers are willing to pay for enhanced experiences (43 percent of Americans subscribing to a top newspaper are willing to spend more on their current subscription, according to Michael J. Wolf's latest Activate report), which explains the popularity of membership programs like "Slate Plus" and "Times Insider" and *The Atlantic*'s "Masthead" program.

The New York Times now has subscribers in 195 countries, which make up roughly 15 percent of its more than 2.6 million paid digital-only news subscriptions and continue to grow at a faster rate than its domestic additions. They've made major international expansion efforts into English markets such as Australia, and they have recently launched a service in Spanish. They realize that in order to reach their ambitious goal of 10 million digital subscribers, they're going to need to target affluent readers overseas. The ability to quickly process a wide range of currencies and payment types is crucial to this growth.

As a student of subscription metrics, of course there's a lot more that I'd like to know. How effective is the *Times* in acquiring new readers? What is the cost of customer acquisition, and what is the monthly churn in the subscriber base? What's the average level of engagement per subscriber? The paper doesn't publicly share a lot of those details, but the bottom line is that today *The New York Times* is looking much more like a smart, recurring revenue-based SaaS platform than a static advertising billboard. *The New York Times* has also been famously circumspect about engaging in content-sharing agreements with Facebook, Google, and Apple, which I think is a good thing. All three are working on mechanisms to incentivize paid subscriptions, but if I were a publisher I would be suspect about any arrangement that entailed handing over my consumer and payment data.

"We are not trying to give away the store here," said Rebecca Grossman-Cohen, *The New York Times*'s vice president of audience and platforms, to *Digiday*. "We know what it takes to build a healthy subscription business and that's building a relationship with readers. To do that, you need to have a direct connection with them. And if we are an atomized set of articles with some light branding that you might recognize or might not recognize as *The New York Times*, it might be harder to do that." Exactly. This is the same reason many publishers rightly rejected Apple's 30 percent "iTunes

Tax"—they weren't happy about the size of the revenue cut, but they were livid about Cupertino keeping all the payment and demographic data. That's giving away the store.

The New York Times should be drowning in VC funding. As *Recode* recently noted, *The New York Times*'s digital paywall business is growing as fast as Facebook and faster than Google. It's following all the standard Silicon Valley best practices: subscription revenue, international expansion efforts, multitiered service offerings, freemium offers, customer behavior insights, and a significant TAM, or total addressable market. "We are, in the simplest terms, a subscription-first business," says *New York Times* CEO Mark Thompson. "Our focus on subscribers sets us apart in crucial ways from many other media organizations. We are not trying to maximize clicks and sell low-margin advertising against them. We are not trying to win a pageviews arms race. We believe that the sounder business strategy for the *Times* is to provide journalism so strong that several million people around the world are willing to pay for it." I hear a lot of echoes of Jeff Bezos in that statement! And for that matter, why aren't more big VC firms flocking to smart, established newspaper companies with a steady base of digital subscriptions instead of these digital journalism ventures that are trying to play ad dollar musical chairs against Google and Facebook? It's a mystery to me.

As the media expert Peter Kreisky once told me, the broader message here is that quality journalism will continue to have solid support from a committed group of readers who care about informed insight and are happy to pay to get it anywhere, anytime (particularly on their phones and tablets). Michael Wolf's latest Activate report notes that if roughly a quarter of the US reading public is paying for news right now, there's at least another quarter of the market that actively seeks out news through password sharing and social media. That's a lot of potential new paid subscribers.

In the case of publications like *The New York Times, Financial Times,* and *The Economist,* as well as consumer magazines like *Motor Trend,* the race is not for scale at all costs but for paid engagement with their audience. That's a clarifying and liberating realization: You don't have to be everything to everyone. You just have to know your reader.

CHAPTER 6

——

SWALLOWING THE FISH: LESSONS FROM THE REBIRTH OF TECH

ADOBE, THE TRAILBLAZER

In November 2011, Adobe's CFO, Mark Garrett, told dozens of Wall Street analysts that he was going to try as hard as he could to make his company's revenue earnings fall as quickly as possible. It was an understandably tense call. Adobe was going to stop selling its enormously profitable Creative Suite software in boxes and move to a digital subscription model: "The faster earnings fall, the better off we are as a company and the better off you are as investors, because millions of people paying us every single month is very compelling from a revenue perspective." The overall revenue wasn't going away, it was simply being pushed out into the future, and Garrett's team took great pains explaining why and how they planned to accomplish that transition.

What precipitated Adobe's decision to move to subscriptions? Its licensed software business in 2011 was a cash cow, generating more than $3.4 billion in revenue at a 97 percent gross margin. Most management teams would be hard-pressed to find much fault with those kinds of numbers. But there were some troubling signs—the business was growing primarily as a result of price increases, and the overall user base wasn't growing. "The number of units we shipped under the old perpetual-licensing model was about three million units a year, and it remained flat for a long time," said David Wadhwani, who was then Adobe's senior vice president of Digital Media, at our *Subscribed* conference in 2014. "We were driving revenue growth by raising our average selling price—either through straight price increases or through moving people up the product ladder."

There were several other red flags. Historically, Adobe had delivered product updates every eighteen to twenty-four months, but they realized that their customers' content-creation requirements were changing much faster than that, with advances in devices, browsers, and mobile apps. They simply weren't moving fast enough. And while everyone had been beaten by the 2008 downturn, Adobe suffered much more than other software companies that had higher margins of recurring revenue—it had no financial buffer. "When we looked at how other software companies were faring during the recession, we saw that companies with high recurring revenue had smaller declines in their growth rates and valuations," said Garrett. "We had a very big drop in both—our revenue dropped about 20 percent, and our valuation fell even more."

Adobe was working harder and harder to make its quarterly numbers. They stepped up their marketing, but they couldn't get the returns they needed. They stepped up their product updates (as much as they could), to little avail. At this point, the user base was actually starting to decline, at the same time that digital publishing—Instagram, online video—was exploding. They were literally and figuratively stuck in the box. The management team was left with two options. One was to essentially treat Creative Suite, which still dominated print publishing, as a bank account for a new line of business that would go after digital publishing. The other option was to double down on Creative Suite and turn Adobe's core franchise into something that could embrace both worlds, which meant continuous innovation, digital services, and lower monthly costs in order to organically increase the user base.

Mark Garrett's pitch for Adobe's shift to subscriptions during that November 2011 meeting was informed and methodical. The next day the stock tanked (though not as much as Adobe's management feared it might).

SOFTWARE'S NUCLEAR WINTER

Today the technology industry is a $3 trillion juggernaut, growing at a healthy 4 percent clip. VC funding is at a decade high, with roughly $84 billion invested in 2017, a one hundred percent increase from 2007. And while those numbers are starting to approach the dizzying highs of the dot-com era, this time the vast majority of the investment volume is driven by the late-stage funding rounds of established companies with solid fundamentals. According to Tomasz Tunguz, a partner at Redpoint Ventures, cash on public software balance sheets has increased by a factor of twenty over the last ten years, and capitalization of the public software companies has increased by a factor of twenty-eight over the same period. There will always be ups and downs (crypto, anyone?), but technology is undoubtedly a growing, vibrant, and increasingly diverse industry.

But it didn't always used to be this way! Ten to fifteen years ago, Adobe wasn't the only software company in the doldrums. Growth across the entire industry was flat or down—the 2001 crash wiped out a decade of gains, and there was talk of a "Software Nuclear Winter." Multibillion-dollar companies like Siebel were being acquired or went out of business. Venture capitalists refused to fund new software start-ups. Tumbleweeds were rolling down Sand Hill Road (metaphorically, anyway). Wall Street declared that the industry had permanently matured and that firms should start paying out dividends like (gasp!) regular old utilities. Financial analysts argued that the software industry should be viewed as an area of "discretionary" spending that businesses would seek to minimize if the economy took another southward turn.

The title of a 2003 *Harvard Business Review* article summed up the prevailing mood at the time: "IT Doesn't Matter." Its author, Nicholas Carr, basically called the entire sector a bunch of glorified plumbers: "By now, the core functions of IT—data storage, data processing, and data transport—have become available and affordable to all. Their very power and presence have begun to transform them from potentially strategic resources into commodity factors of production. They are becoming costs of doing business that must be paid by all but provide distinction to none." Consolidation was rife, as Oracle swallowed up struggling companies in order to maintain market share through sheer inertia. After laughing off the iPhone in 2007, Steve Ballmer's Microsoft was stuck firmly in the midst of a ten-year rut.

While the big tech dinosaurs were limping through this extinction event, a nimbler breed of SaaS companies was coming onto the scene, but there were a lot of doubts about them. They only worked for small businesses. Their "dial-up software" couldn't be configured or integrated into larger systems. They were a fad that appealed only to cash-strapped companies in decline. And who would trust their data on someone else's servers? But the mammals kept growing and growing, with Salesforce leading the way, and then another big asteroid hit: the 2008 crash. Suddenly these

SaaS alternatives started looking a lot more appealing to companies weathering the economic downturn, and the prevailing attitude went from "We'll run some less important systems on the cloud" to "We're a cloud-first shop." These days, more and more CIOs are staring down the ridiculously huge upgrade bills for their on-premise ERP systems, and they're wondering why they should even bother when there are so many cheaper, smarter SaaS alternatives on the market.

These days, everyone understands that subscriptions are the dominant business model for the technology industry. Gartner predicts that by 2020, more than 80 percent of software providers will have shifted to subscription-based business models. As a recent Deloitte paper notes, the big technology firms simply can't afford not to offer subscription models: "As more and more customers demand more flexible payment models, the continued viability of many companies, and even entire industries, is being threatened. Those that fail to at least explore consumption-based offerings may end up on the path to obsolescence." Across the board, perpetual license and maintenance revenues are slowing or in decline. There's no growth left in on-premise software, and a lot of younger SaaS companies founded over the last ten years are starting to see real gains in market share. Hardware is shifting as well—the success of Amazon Web Services has convinced IT buyers to shift from big, expensive capex-based installations to opex-based rental agreements. The big firms are struggling to catch up because they can see where the market is heading.

Today's innovative companies are increasingly pursuing recurring revenue–based business models and are relegating their ERP systems to a commodity general ledger in the process. The old, cumbersome "one size fits all" ERP model increasingly means that everything gets billed to Oracle, and nothing gets done particularly well. From customer service to expenses to forecasting to billing, more and more companies are taking advantage of "plug and play" SaaS providers that are maniacally focused on performing one core function really, really well. Concur does nothing but think about travel expenses all day. Salesforce is synonymous with CRM. Aviso is putting the latest machine learning technology into sales forecasts. My company, Zuora, lives and breathes subscription billing, commerce, and finance. These SaaS solutions are way more nimble and effective than a massive Oracle installation.

So how can the dinosaurs survive? To figure that out, we're going to need to talk about fish.

THE FISH MODEL

To understand why traditional software companies were facing such a steep uphill climb, you have to understand the Fish Model. In their excellent book *Technology-as-a-Service Playbook: How to Grow a Profitable Subscription Business*, Thomas Lah and J. B. Wood refer to this transition period as "swallowing the fish," as the revenue curve temporarily dips below the operating expense curve before climbing back upward again:

> *The fish is what happens when a traditional company starts to shift its revenue mix from an asset purchase model to a subscription model. In this scenario, the company experiences a string of quarters where top-line revenues shrink as revenues from large, pay-up-front deals are replaced by recurring subscriptions without the big up-front payment. At the same time as revenues dip, the company must make investments in many of the new capabilities and structures that are required for profitable XaaS. The traditionally profitable and stable mix of more revenue than costs on the left side of the chart is replaced with a tumultuous period of costs exceeding revenue.*

So, a fish:

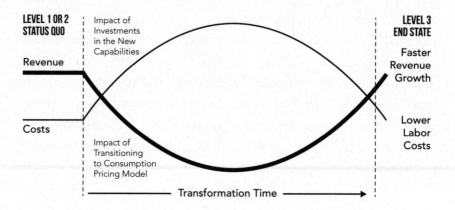

As Lah and Wood note, management teams chasing after quarterly numbers generally don't like the look of that fish. They would just as soon avoid it altogether. There are boards and investors to consider, not to mention the fact that as traditional unit-based companies, Wall Street measures them on strict GAAP-realized profits, not growth rates based on deferred revenue. When Salesforce, for example, reports its annual revenue figures and then notes that its deferred revenue figures are 60 percent of that number, that means it gets to start next year with over half its revenue target in the bank. Wall Street appropriately gives them much higher multiples than it does a company that can't report that second figure. But those management teams are trapped in the old GAAP rules, and they wind up getting stuck defending an outdated model: "Profitable, incumbent players seem to stand still as new entrants disrupt a marketplace. They are reticent to disrupt their profitable economic engine—even as customers start to leave and revenues start to shrink."

This was the challenge the Adobe management team was facing in 2011. They knew they had a big fish to swallow.

So how did they do it? Well, once the executive team decided to pursue this model, it committed itself to relentlessly overcommunicating. The idea was to drown controversy with transparency, but they needed slightly different messages for each of their core constituencies. They started with their employees. There were some pretty serious organizational changes that had to happen. If some long-term employees were freaking out, you could certainly empathize. The finance team had to move from simple transactional sales to suddenly billing three to four million individuals every single month. The product team had to move from yearly to monthly feature updates, as well as tackle new challenges like uptime, disaster recovery, and security. Management put together a jokey self-help video for the sales team called "revenue addicts," whose goal was to move the mindset (and commission structure) from quarterly numbers to long-term bookings.

Which brings us back to that infamous analyst meeting in New York. In 2011 there was little precedent for an established software company making a successful transition to subscriptions, due to organizational inertia, market myopia, system constraints, or a static product mindset (or all of the above). Most management teams were paralyzed by the same predictable set of questions: If we take revenue over time versus up front, doesn't that hurt our bottom line? Won't subscriptions lower our margins? How do we get our sales team to sell this stuff? How will investors react? At the time, most of the success stories involving digital subscriptions involved young SaaS companies taking market share from older companies desperately trying to prop up mediocre but profitable business units.

Adobe faced two significant challenges at that meeting—they not only had to convince Wall Street of their vision, but they had to ask them to rewrite their financial models in order to accommodate it. Thirty years of looking at Adobe in terms of unit sale analysis was about to go out the window. Garrett was arguing that Wall

Street needed to embrace an entirely new set of metrics that had no disclosure requirements or bearing on a company's GAAP financials. To make that argument, Adobe shared much more financial information than it had in the past, as well as a clear set of target "markers" of future subscriber and annual recurring revenue (ARR) growth numbers. As noted, the results were mixed (at one point NASDAQ called to ask if they wanted to stop trading), but the management team stuck to the plan, knowing that they next had to approach their most important constituent of all—their customers.

Here the Adobe team made another smart decision—they gave their customers some time. While promoting digital subscriptions, they also launched Creative Suite under the traditional perpetual-licensing model in May 2012, giving themselves a year of parallel sales models. Even when they switched to digital subscriptions in May 2013, they didn't pull the old perpetual product completely—they just took pains to let their customer base know that they would no longer be updating it. Wadhwani delivered a two-hour keynote at Adobe's annual MAX creativity conference, making a frank pitch to his customer base: acknowledging concerns, walking through the company's reasons, and explaining the benefits. This was followed up with a sustained evangelist outreach effort that featured face-to-face meetings with more than fifty thousand customers.

In three years, Adobe Creative Cloud went from almost no subscription revenue to a virtually 100 percent subscription model. Today Adobe's transition to digital subscriptions is taught in business schools. Adobe provided the "textbook" case that inspired Microsoft, Autodesk, Intuit, and PTC. When Adobe announced its transition, its stock was trading at around $25. Its income fell by almost 35 percent the following year. Today Adobe's stock is trading at over $190, it's growing at 25 percent a year, and it has roughly $5 billion in ARR (up from practically none in 2011). Over 70 percent of its total revenue is recurring. Amazing.

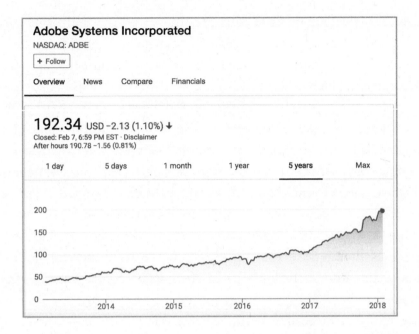

PTC PIVOTS

Today the entire software industry is following Adobe's lead. In February 2015, for example, Autodesk announced the transition from perpetual licensing to pay-as-you-go subscription plans. By August 2016, Autodesk's stock jumped to an all-time high following record increases in cloud subscriptions as their "customers and partners embrace a model that has greater flexibility and a better user experience," according to Autodesk's then-CEO Carl Bass. And why did Microsoft, which has been public since 1986, see its stock hit an all-time high in July 2017? Because it rode the shift to become a successful SaaS company with its "Commercial Cloud" business nearing its fiscal 2018 goal of a $20 billion annualized revenue run rate and its Office 365 Commercial business beating out its traditional licensing business in revenue generation. There

are countless examples of software companies that have successfully made the shift to subscriptions and subsequently driven higher
valuations and more shareholder value: IBM, Symantec, Sage, HP
Enterprise, Qlik.

Another big reason? IT buyers prefer opex to capex. Historically
software companies have preferred capital expenditures (capex) for
technology investments, as this afforded them the ability to take advantage of amortization and depreciation of the capital investments
over a period of time. But as technology shifts to the cloud, there's a
complementary shift happening in favor of opex over capex. Operating expenses bear the advantage of a pay-as-you-go model for services used with comparatively little to no up-front investment. Not
only is this a greater value prop, as a business is getting exactly what
it pays for, but it's also a strategy for freeing up cash to drive growth,
and a way for large enterprises to be nimble rather than locked into
expensive IT infrastructure that lacks flexibility and often serves as
nothing more than a bottleneck for transformation.

All this being said, some companies are managing this transition better than others. PTC, for example, is killing it, so let's take
a look at their story. PTC (Parametric Technology Corporation) is
one of the fifty biggest software companies in the world. Its customers design aircraft, plan buildings, make sneakers, build tools,
pioneer new medical diagnostic technology, and much more. A few
years ago, PTC's earnings took a dip (sound familiar?). In the second quarter of 2015, PTC recorded $303 million in revenue. A little
over a year later, that number dropped to $288 million. In the same
period, earnings swung from a $17.4 million profit to a loss of $28.5
million. But as I write, PTC's stock is up 135 percent over less than
two years, from a low of $28 in February 2016. In less than two
years it has added more than $4 billion in shareholder value.

Just a few years ago, PTC found itself in a similar position to
where Adobe was in 2011—chugging along, like the rest of the tra-

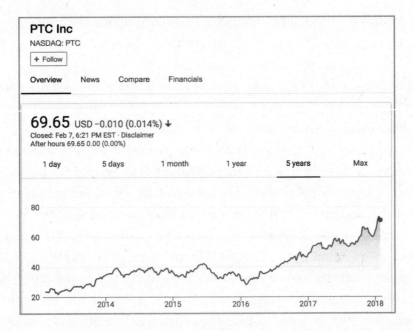

ditional software sector, at low-single-digit growth. PTC had to begin every financial year at zero—revenue had to be clawed together one deal at a time, only to vanish again in twelve months. Overall revenue growth was a nice idea, but not at the cost of quarterly profits—unwittingly or not, PTC had lashed itself to a strict, GAAP-based valuation model. As a result, many wondered if it was simply a mature company that should be passing its maintenance revenue to investors in the form of dividends—PTC's valuation multiples were stuck in the 1x–3x price/sales range.

At the same time, PTC found through its own survey analysis that more than 90 percent of its customers shared a desire for subscription-based pricing—perhaps this was the "Adobe effect" at work. These were creative professionals who liked the fact that they could get a quick approval from their finance department for an easy monthly opex spend versus a nightmare capex project. They could justify ROI to their bosses ("This is what I used, so this is what I spent"), which is usually a much more difficult exercise with million-dollar project spend. There was way less bureaucracy and no painful IT integrations involved. So PTC announced a broad, systemic shift from perpetual licenses to cloud-based subscriptions, and it also confidently predicted that this shift would rekindle growth, expand margins, and maximize long-term shareholder value. It wound up going three for three.

In October 2015, at the start of the journey, PTC told investors and analysts that in five years (FY2021) it was aiming for $1.6 billion in revenue, 10 percent revenue growth, an operating margin in the low thirties, and 70 percent of bookings coming from subscriptions. In stark contrast to Adobe's announcement, PTC's analyst call resulted in a nice little pop for them—PTC went from $32 per share at the end of September 2015 to about $37 per share at the beginning of November, or a 15 percent increase in its valuation. But just one year later, the news was significantly better. When PTC gave an update on its transition, it moved its FY2021 targets up by a full year. It also raised its target for pure subscription bookings from 70 percent to 85 percent. This was on the heels of FY2016 earnings results that consistently demonstrated the transition was tracking ahead of PTC's initial plan.

Let me just compare those two analyst calls again: In 2015 PTC predicted $1.6 billion in revenue with 10 percent sustainable growth in FY2021, a steady-state subscription mix of 70 percent, non-GAAP operating margins in the low 30s (from the mid-20s), and $450 million of free cash flow. In 2016 it revised those estimates to $1.8 billion in FY2021, revenue growing at a sustainable 10 percent–plus growth rate, 85 percent steady-state subscription mix resulting in 95 percent of their software revenue being recurring, non-GAAP operating margins in the low 30s, and free cash flow of $525 million. Wow! PTC has clearly swallowed the fish, the same way Adobe did. It's reached the fabled inflection point where recurring revenues outgrow recurring costs, after which the unit economics favor hockey-stick growth figures. PTC's subscription ACV (annual contract value) guidance at the beginning of fiscal year 2016 was $43 million. It delivered $114 million, almost three times that original target.

PTC saw the shift toward subscriptions coming and reacted with a smart, emphatic transformation. Its management team knew full well that the subscription model creates deferred revenue, so quarterly GAAP metrics can take a short-term hit. While this dynamic has caught other teams flat-footed, PTC embraced it and kept the public investment community informed every step of the way. As a result, PTC is driving growth, showing substantial margin improvements, and discovering whole new territories of shareholder value.

Here I'd like to call out two key points from an *HBR* piece called "How Investors React When Companies Announce They're Moving to a SaaS Business Model": (1) Like Adobe, you don't have to go all in, all at once. The study found that investors increase their valuations of the software vendor's stock by an average of 2.2 percent if the vendor makes clear in its announcement that the SaaS offering is provided in parallel to a perpetual licensing model. (2) You don't have to do it all yourself. Announcements that implied the SaaS offering would be built in cooperation with cloud infrastructure and platform providers increased company valuations by an average of 2.9 percent.

THE HARDWARE SHIFT: CISCO

It's not just software companies making the shift. Hardware companies are adopting subscription models in droves. Take Cisco, which sells the routers and switches that forward data packets between networks. Most of the internet runs on Cisco hardware. Cisco's business used to be pretty straightforward—it sold tons of data equipment to thousands of companies for lots of money. But four or five years ago it was facing some serious headwinds—thanks to cloud computing, its clients didn't need as much of its hardware. All those in-house data centers were ascending into the cloud.

Ben Thompson of *Stratechery* sums up the appeal of cloud computing quite neatly: "The true value in a public cloud isn't necessarily up-front cost-savings . . . but rather the value of the optionality that comes from building on an infinitely scalable and malleable infrastructure that is pay-as-you-go, as opposed to a massive capital investment that, if things don't work out, quickly becomes a millstone around your neck." Cisco was at risk of selling companies thousands of new millstones. And those millstones were already at risk of becoming commodified thanks to cheaper competitors and software-based alternatives. While its hardware business was flat, management saw that most of its growth was coming from its new service acquisitions around security and collaboration. It was buying growth.

If you were in the railroad industry, would you be more interested in the business of laying the tracks or delivering the freight? One element is discrete and transactional (how many new rail lines do you really need?); the other represents ongoing value. A new management team at Cisco decided to go all-in on services, which by definition meant subscriptions. But how do you sell routers and switches on a subscription basis? By focusing on the data inside all that hardware—the freight, not the tracks. Cisco's latest set of Catalyst hardware comes embedded with machine learning and an analytics software platform that helps companies solve huge inefficiencies by reducing network provisioning times, preventing security breaches, and minimizing operating expenses.

"I have options to talk to my customers about operating expenses and taking the lumpiness out of tech refreshes," says Mike

Girouard, vice president of sales at TekLinks, a Cisco partner. "It makes hardware more affordable from a capex perspective and operationalizes the software. It makes our cash flow less lumpy, and it does the same for the client. You can see Cisco moving to a pay-for-what-you-use approach. It's more natural. It spreads the risk out for customers and for us, and it forces Cisco and us to have a tighter relationship with customers."

Cisco isn't just managing a dependable if relatively flat hardware business while it hunts for growth in software and services. It's embracing subscriptions in a broad, systemic way in order to shift from selling boxes to selling outcomes. Its new cloud-based management services help mitigate the boom-and-bust effects of new product cycles. It doesn't have to act like a retailer chasing after make-or-break holiday sales in order to make its annual number. Today almost a third of its revenue is recurring, which is resulting (as CFO Kelly Kramer is quite happy to point out) in a

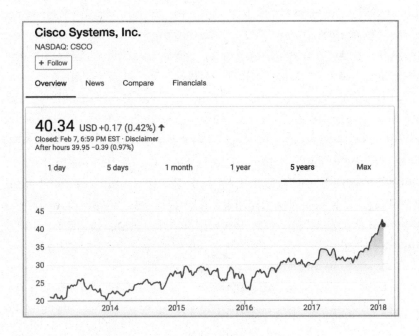

short-term hit to its GAAP revenue numbers. Again, *standard revenue loss is a good thing.* That's a sign that you are carrying your book of business out into the future. Cisco is swallowing the fish.

Give credit where credit is due: Adobe blazed the trail. That infamous analyst call back in November 2011 was a turning point in modern business history. Today the market is much more sophisticated about subscription models as growth engines. What do the Adobe and PTC transition stories have in common? As Lah and Wood note, successful enterprise transitions share some common themes: "They communicated their transition objectives very clearly and they committed to a firm timeline. They ran with an aggressive profit maximizer mentality. They were open and transparent in their financial reporting . . . the louder and more publicly aggressive companies seem to be about their pivot to technology as a service, the better chance their stock price has of weathering the transition period."

Today the tech sector is booming again. According to the St. Louis Fed, since the Great Recession ended, jobs in the tech sector have been growing by 20 percent, compared with 11 percent growth in the rest of the private sector. In 2015, tech sector employment exceeded 4.6 million workers, passing its 2000 peak. I don't think that was a given. I think Nicholas Carr could have easily been right—thanks to rampant commodification, the tech industry could have been relegated to a steady if underwhelming sector of the global economy. Until 1996 most of the jobs in the tech industry were in manufacturing—you went to work at a fabrication plant, not a hip start-up office. Today nearly 80 percent of tech workers are in services. When an industry embraces this shift, it finds growth.

CHAPTER 7

IOT AND THE FALL AND RISE OF MANUFACTURING

During the early days of Zuora, we would often sit around a dinner table with a bottle of wine and try to stump ourselves—what *can't* you subscribe to? Media and software moving to subscriptions was a no-brainer, but that was all just ones and zeros. What about the really *heavy* stuff—the buildings, the industrial equipment, the construction supplies? How would you subscribe to a refrigerator, for example? How would you subscribe to a floor? How about an excavator? What about a roof? We eventually wound up supporting subscriptions for all of these things, but early on, they were just fun theoretical discussions. Here's the secret we used to answer all of them in the affirmative—*tease out the service-level agreement that sits behind the product.* It works for everything. So instead of a refrigerator, it's the guarantee of fresh, cold food. Instead of a roof, maybe it's a guaranteed source of solar energy. Instead of excavators, it's the expeditious removal of a certain amount of dirt.

Today all those after-dinner conversations are becoming a

reality, particularly with regard to dirt, heavy equipment, and service-level agreements. Let's take a look at construction. One of the first steps in building something is figuring out how much dirt you need to dig in order to lay the foundation. Work site surveying is a pretty inefficient process. Manual surveys generally have a 20 to 30 percent margin of error—that affects equipment rentals, material purchases, labor hiring, completion plans, everything. According to McKinsey, big construction projects routinely run up to 80 percent over budget, and they typically take 20 percent longer to finish than the original completion plan. Surveys can also take several weeks to complete. Plus, information is routinely scattered across various blueprints and databases, making it much easier for mistakes to happen. All this is about to change.

Today Komatsu can finish a work site survey in thirty minutes. Komatsu was founded in 1921 and is one of the world's oldest construction and mining equipment manufacturers. A couple of years ago, it launched a new service called Smart Construction, which takes advantage of the same kind of new radar technology that's driving automobile automation in order to take the manual guesswork out of site surveying. When the Komatsu team arrives at your job site, the first thing they do is unleash a bunch of awesome-looking drones into the sky (they have some great videos on YouTube). The drones create a 3D-rendered topographical model of your work site in centimeter-level detail. Komatsu maps that 3D rendering to your work site blueprint in order to calculate the exact area and volume of earth that needs to be removed. Then, much like the artificial intelligence programs that people play chess against, they run thousands of simulations of possible scenarios within this new virtual work site in order to determine the best possible approach. The result is a finished project plan, with materials, equipment, labor, and a work schedule detailed down to the last hour.

The labor part is particularly interesting. At home in Japan,

Komatsu is dealing with an aging workforce. Here in the United States, there's strong manufacturing demand but a lack of qualified equipment managers. So Komatsu feeds your project plan into its fleet of semiautonomous excavators, bulldozers, and backhoes, and these giant robots basically take care of the project for you. In much the same way that the pilot of a 747 only "flies" their aircraft for seven to ten minutes per flight, your equipment managers are mostly there to supervise. And like something out of *Star Wars*, site managers can sit in front of their 3D virtual work site, follow progress in real time (rotate, zoom in, zoom out, etc.), and run simulations for any plan changes they might want to consider. Amazing.

So what's my company doing with Komatsu? The same thing we're doing with Caterpillar—we're helping them change the question from "How many trucks can I sell you?" to "How much dirt do you need moved?" By handling the subscription finances behind these services, we're helping to power dirt removal as a service. Caterpillar recently came to our *Subscribed* conference in San Francisco to talk about how they're tackling the challenge of getting involved in customers' businesses even before there is a work site. They're asking big questions around how they can analyze data from one project to another and even help customers win more business. Case in point: A couple of years ago Caterpillar was approached by a client with more than sixteen thousand pieces of equipment. They wanted to be able to manage every single one of those assets from a single screen: utilization, fuel amounts, idle time, etc. Caterpillar retrofitted the entire fleet, and a year later the client reported an almost 20 percent utilization increase.

Caterpillar also manufactures those giant mining trucks. I'm not sure if you've seen pictures of them, but they are ridiculously huge— the truck bed can fit more than two hundred regular-sized cars. The driver sits two and a half stories up in the air. They look like giant Tonka trucks, but they're essentially semiautonomous rolling factories. A Caterpillar client had one break down in the field, and it cost them $650,000 and nine hundred hours of downtime. Today Caterpillar offers an analytics platform called Cat Connect Solutions that lets site managers avoid those kinds of problems—or at least anticipate them. They can figure out when a machine needs servicing by monitoring things like vibration patterns and comparing them with the past usage data of identical machines. That helps them schedule at-risk machines into a maintenance session that could bring those numbers down to $12,000 and twenty-four hours of downtime. Now, that's an incredibly valuable service. And where did it come from? The information generated by Caterpillar's fleet.

Manufacturers and original equipment manufacturers (OEMs) all over the world are waking up to the fact that there are dozens of potentially new value-added services that are sitting right inside their servers—let me explain what I mean by that.

THE BIGGEST TRANSFORMATION OF ALL

If the context of this book is the sweeping shift from products to services in retail, transportation, media, and technology, as well as the changes in business mindsets that this shift requires, then the industry that stands to benefit the most from that transformation is manufacturing. That may sound surprising. We've all been told that the global economy is shifting away from hard goods. Manufacturing as a percentage of the US GDP peaked in the 1950s, and manufacturing employment has been in steady decline since the late nineties. According to the Bureau of Labor Statistics, today overall employment in manufacturing in the United States is at the same level as it was before the country entered the Second World War. Almost a third of the workforce was building things in factories in the early fifties—today it's a little under 9 percent. Around the world, the picture isn't much better. According to the IMF, the drop in global productivity after the Great Recession has been "widespread and persistent across advanced, emerging, and low-income countries." The world is running out of growth.

But here's the thing—the manufacturing industry is still enormous. According to the National Association of Manufacturers, manufacturers contributed $2.2 trillion to the US economy in 2016, or almost 12 percent of the GDP. There are roughly 12.5 million manufacturing workers in the United States. Since the Great Recession, manufacturers have hired more than one million US workers. Considered on its own, manufacturing in the United States would constitute the ninth-largest economy in the world. So manufacturing is huge—way bigger than many of us may realize.

But why is manufacturing about to get much bigger, much faster, than any other sector I've talked about in this book? Because we're on the verge of a manufacturing revolution that is very quickly going to result in double-digit gains in global productivity, output, and growth. Manufacturing is the old man who wakes up to find himself young again—it's a story as old as Gilgamesh. As LL Cool J says: "Don't call it a comeback. I've been here for years."

THE INTERNET OF THINGS

Over the past five years, thousands of manufacturing businesses around the world have quietly been investing huge sums of money into sensors and connectivity. They've been hard at work putting sensors into everything they make: doors, chairs, pipes, tiles, windows, tables, sidewalks, rebar, lights, shoes, bottles, tires, bricks, etc. According to various predictions, by 2020 we're expected to have more than a billion smart meters, 100 million connected lightbulbs, more than 150 million 4G-connected cars, 200 million smart home units, several billion smart clothing units, more than 90 million wearables. And what do these sensors allow these products to do? Collect and transmit data—lots of it. All of these products will be beaming information back into centralized servers, so companies can start using analytic platforms to look for patterns and ways to improve things (you want to talk big data? This is BIG DATA). This entire ecosystem is commonly known as the Internet of Things, or IoT.

IoT is the digitization of the physical world through sensors and connectivity. That may sound kind of jargony, but it's accurate. When you digitize something, you convert it to data in the form of numerical digits, which allows it to detect, communicate, and respond to other digitized objects. Eventually, all the manufactured objects on the planet will be able to receive and transmit data. Analytic services based on all the data generated by these billions of connected objects will result in system efficiencies and business ideas worth trillions of dollars. Right now we're in the first "efficiency" stage of IoT, which is about diagnostic systems improving efficiency and productivity. But things are about to get much more interesting—IoT is moving beyond efficiencies and into possibilities. Here's a compelling quote from Scott Pezza of Blue Hill Research on the topic:

> *If you currently sell products that collect some sort of data (or could be retrofitted to do so) and there is someone out in the world who would find that data valuable, IoT is a new revenue source for you. If you sell physical products that degrade or need to be serviced, IoT means you can offer remote monitoring services, or preventative maintenance services—new revenue streams. In the alternative, you can increase the attractiveness (and value) of those products by giving customers the ability to conduct that monitoring and maintenance themselves. If you sell services that could be expanded if you only had access to more data, it's new money. And if you sell tech-*

nology to help sense conditions, facilitate secure communications, conduct analysis, manage service provisioning and billing, or forecast and plan revenue—this market is going to need you.

By 2030 IoT is projected to explode into a sector that will be roughly the size of the current economy of China, or around $14 trillion. That figure represents about 11 percent of the global economy. Those are staggering numbers, but I actually think they're understated. Right now there are roughly 250,000 manufacturing firms in the United States involved with motor vehicles, aerospace equipment, robotics, fabricated metals, electronics, plastics. And, of course, there are millions more all over the world. All of them are turning into IoT companies. When everything has a digital component, then the old distinction between "heavy" industrial equipment and "light" digital services becomes immaterial—literally.

This year the entire global industrial output, or the systems, factories, and labor force *behind everything we make*, is expected to enjoy a double-digit increase in delivery and supply chain performance, primarily thanks to IoT. All this connectivity is set to ignite a revolution in manufacturing, and it's sorely needed. As industrial systems thinker Olivier Scalabre pointed out in a great 2016 TED talk, every sustained period of global economic growth over the last 150 years has been instigated by a manufacturing innovation—the steam train in the nineteenth century, the era of mass production at the start of the twentieth century, and the first wave of factory automation that began in the 1970s. But today we are emerging from a long period of stagnation. As Scalabre notes:

It's not as if we've done nothing since the last manufacturing revolution. Actually we've made some pretty lame attempts to try to revitalize it. For example, we've tried to relocate our factories offshore in order to reduce cost and take advantage of cheap labor. Not only did this not inspire productivity, but it only saved money for a short period of time, because cheap labor didn't stay cheap for long. Then we tried to make our factories larger, and we specialized them by product. The idea was that we can make a lot of one product and stockpile it to be sold with demand. Today . . . most of our factories look the same as they did 50 years ago.

That's all about to change.

DIGITAL TWINS

You may have grown up thinking of General Electric as a kitchen appliance company. Today they build wind turbines, jet engines, oil rigs. They also have a thriving data services business—maybe you've seen some of their commercials aimed at recruiting more developers. They have over $3 trillion in assets that they manage on a regular basis, and today almost all of them have twins—more specifically, digital twins. We recently hosted Gytis Barzdukas, vice president at General Electric Digital, at our *Subscribed* conference in San Francisco. He pointed out that digital twins don't just represent how their physical assets were designed or how they were built—*they display how those assets are operating in real time.* A jet engine that's being operated in the US Southwest, for example, has a different digital twin from one that primarily flies across the North Sea. Over time, those engines behave and degrade in different ways, and they transmit usage data accordingly. Very soon, engineers on the ground will use augmented reality headsets to see all this information overlaid on the jet engines when they inspect them. The digital twins will point out wear and trouble spots and offer opinions on how to resolve issues based on asset history.

Essentially, GE operates its own social network for heavy industrial machinery. It's sort of like all these power grids and oil refineries and MRI machines have their own Instagram accounts, but instead of pictures of beaches or food, they're sharing fuel consumption, hydraulic pressure, usage hours, decay rates. "First there was the consumer internet, and then the enterprise internet," as Barzdukas said, "and now we're moving into the third generation: the industrial internet. It's not just about having our phones connected or our enterprise applications connected and operating on subscriptions models. Now it's the big machines." So far GE has built more than 600,000 of these digital twins. And just as social networks changed our world, this third-generation industrial internet is going to transform manufacturing.

As Barzdukas pointed out, it's relatively simple to retrofit existing machinery with sensors. A typical consumer phone has between twelve and fourteen sensors in it that handle everything from lighting to synaptics. Lots of industrial machinery, of course, already has sensors, but right now the connectivity enabled by those sensors is the equivalent of trying to communicate with two tin cans and a piece of string. A typical oil rig, for instance, has thirty thousand sensors, but usually only 1 percent of the information generated by those sensors is examined in any meaningful way because those sensors are fairly single-minded—they're just trying to detect individual anomalies, not optimize the system as a whole or peer around corners to make predictions. So sensor retrofitting and networking, or the "implementation phase" of IoT, are set to become a huge growth industry in the years ahead.

And what happens when you have a vast network of digital twins that represent every asset across your entire product line? Well, the first beneficiary was GE itself. What if you had one engine acting up, or one compressor behaving strangely, among thousands? Imagine that if instead of trying to catch problems with expensive and laborious mass maintenance procedures, you had a network of digital twins sending you relevant signals from individual assets. Well, you could fix the biggest problems much faster. GE quickly realized more than $200 million a year in savings, simply by improving their efficiency. Then they turned their new platform into a stand-alone service called Predix, which is an ecosystem of applications with shared data that work together to improve performance and increase efficiency.

But IoT isn't just about efficiency and diagnostics. IoT is allowing manufacturing companies to reimagine their businesses in really profound ways.

FROM PRODUCTS TO OUTCOMES

All this new connectivity and network intelligence means that more and more manufacturers are beginning at the end. What do I mean by that? These days the leaders in manufacturing are partnering with clients to come up with an optimal end state—a health care IT firm signs a contract with a hospital promising to reduce its patient readmittance rate by a certain percentage, for example, or SpaceX signs an agreement with NASA guaranteeing a particular cabin environment to transport lab rats to the International Space Station—and then compiles the technology and resources to make that outcome a reality. Hive, a UK-based home security company, starts with a statement like "Our customers never want to come home to a dark house" and then assembles the technology and services to make it happen. Service-level agreements are replacing bills of sale.

What these companies are realizing is that IoT enables them to view their products as whole systems, as opposed to individual units that are sold to strangers. And these systems constitute a core competitive advantage. They allow them to give their customers what they actually want—the outcome, not the product. The milk, not the cow. That's the story of Komatsu and Caterpillar in the construction industry. And this is why we believe that every company has the potential to reinvent and thrive in the Subscription Economy.

Every day brings dozens of new consumer IoT stories—some kind of silly, some truly compelling. The FDA approves an EKG sensor for the Apple Watch that will help push health care toward remote monitoring rather than periodic testing, potentially saving billions of dollars in health care costs. Whirlpool's new ovens know how to scan recipes and cook meals accordingly. The Disney "magic band" lets you buy food, reserve rides, check in to your hotel room, and unlock special "surprises" around its theme parks. Sensors in Johnnie Walker Blue Label whisky bottles let people know if the bottle has been tampered with, or where it sits in a supply chain. Sensors in football helmets help team physicians keep track of potential head trauma, and inertial sensors in bats and baseballs help teams improve player mechanics. A company called Chrono Therapeutics helps patients administer medication through skin patches that automatically handle dosage and timing issues. Thync is a subscription-based wearable that uses safe, low-level electrical stimulation to help you relax, improve your mood, and sleep better without chemicals. The physical world is starting to "wake up."

The IoT stories are just as fascinating on the industrial side, because they're tackling the same issues that our own smart home technology deals with (security, lighting, sound, temperature, utility consumption), just at a massive scale. Today about half the world's population lives in a city—that number will approach 70 percent by 2050 and will constitute a total urban environment roughly the size of Australia. It's imperative that these places become sustainable environments for healthy lives. Everything is getting optimized: emergency response routes, waste collection logistics, air quality, traffic congestion, public energy usage. Today Barcelona saves itself $37 million a year just on its lighting grid, and its IoT initiatives have resulted in more than forty thousand new jobs. Every manufactured environment on the planet is turning into a data-driven test bed.

Suddenly, buildings are starting to talk to each other. FloorIn-Motion from France delivers connected floors that monitor human traffic patterns in order to manage building energy usage, and, in the case of medical facilities, delivers real-time alerts when patients or senior care residents appear to be having difficulty navigating rooms or hallways. Schneider Electric and its partners are working with farms in New Zealand to retrofit their irrigation systems with sensors that monitor their water usage down to the liter, as well as weather patterns (e.g., "Do I need to water at all?") and favorable spot electricity pricing (e.g., "How can I buy and store grid electricity while it's cheap?"). Supply chains are starting to cognify and manage themselves. Honeywell is working with Intel to let logistics managers not only track packages of sensitive electronic equipment, but also monitor their "health" via a simple stick-on sensor that tracks location, shock and tilt, light, humidity, temperature, and potential tampering. That allows for detailed, real-time inventory tracking by land, sea, or air. The stories go on and on, but they all share a common through-line: connectivity turns products into services, which allows businesses to build around outcomes, not assets.

Another great example of the kind of outcome-based mindset enabled by IoT is Arrow Electronics. It started in 1935 selling radio sets, a relatively new retail technology at the time. Today it's a $24 billion business operating in fifty-six countries. Arrow has roughly eighteen thousand employees that include thousands of field application engineers and systems engineers who help companies bring all sorts of cool new technologies to scale. For many decades, Arrow was happy to conduct its business as a straightforward and extremely profitable parts warehouse. It distributed electronic components to thousands of technology companies.

"Before, you would come in and say I want a processor or I want an electromechanical component; it could be a crystal oscillator or some kind of passive device. You would have a conversation with us about that, and we would go and buy that from someone and sell it to you for a slight markup," said chief digital transformation officer and president Matt Anderson. But Anderson said the company recognized a decade ago that they needed to shift their business model from a commodity provider toward value-added services. They needed to act less like a Home Depot for the tech industry and more like a Bell Labs. Now they're having very different conversations with their customers.

Today they're working with a major fast-food dining chain to put sensors on every single piece of kitchen equipment in order to create a "responsive" menu that can better handle surge demands. They're helping farms build networks of insect pheromone detectors that automatically spray dispersants rather than using pesticides, leading to healthier food. They're even working to build a better *mousetrap*: an automated trap that monitors activity and effectiveness, certifies that it's meeting regulations that restaurant chains and grain elevators must abide by, and notifies the manager of any problems. All that innovation probably won't make much of a difference to the mouse, but it's still pretty impressive.

One question that we get a lot is: How do I sell this stuff? The data inherent in a connected device means you can sell the same information to several different kinds of customers: consumers, advertisers, resellers, industry groups, etc. Multiple beneficiaries give you lots more flexibility in terms of pricing and packaging. Again, let's head to Sweden for some more enlightened thinking around creative usage and pricing strategies. Based in Uppsala, Ngenic sells a smart thermostat, or an "extra brain used to heat up your house." Ngenic offers three basic purchase plans: buy the thermostat outright, buy it for a lower price with a small monthly subscription, or buy it as part of a discounted bundle with an energy provider. More than half of Ngenic customers wind up saving more than 10 percent of their heating costs over the course of a year. The device monitors their home energy usage by taking into account variables like sunlight, occupancy, and (this is where it gets really interesting) favorable electricity rates.

Unlike here in the United States, where you generally get assigned a single power utility, Sweden has an extraordinarily deregulated energy market. There are more than 120 electricity providers, and most customers have rate agreements that can change on an hourly basis relative to supply and demand. Ngenic's physical asset is relatively inexpensive to manufacture and assemble. And thanks to all this rampant free market activity in supposedly socialist Sweden (most kilowatt-hours are bought and sold at least six times before reaching the end customer), electricity there is really cheap. So how does Ngenic create value? It sells to consumers to help them save energy and protect the environment. It partners with suppliers to add "green intelligence" and differentiation to their value proposition. Its devices conduct arbitrage with wholesalers (sometimes on a minute-by-minute basis) to buy more electricity when it's cheaper and less when it's more expensive.

The lesson here is that in IoT, there's no such thing as a straightforward B-to-B or B-to-C market approach. You can have many different kinds of customers. The physical device is just an enabler. Your value lies in your IP, the usage data from your customer base, and your ability to trade information across multiple markets. As Ngenic's CEO Björn Berg told us, "You have to think a little bit like Google and search. Nobody pays for the search itself. But everybody knows that Google profits on you searching. And the benefits you receive are worth more than the information you're giving away. That's the model you have to pursue. What's the value I can create from the information generated by my connected device? That's where you should focus." Ngenic is helping itself (and others) succeed by providing value in a crowded market and offering different pricing and packaging bundles to different kinds of clients. That's a lesson for all of us.

THE FUTURE OF MANUFACTURING

When you take the raw data generated by all these millions of digital twins and interpret that data with analytic software, then you can sell that new intelligence as a *service* that can become as valuable as having electricity, Wi-Fi, or running water in your house. Another word for these kinds of analytic services, of course, is artificial intelligence. Kevin Kelly, in his great book *The Inevitable: Understanding the 12 Technological Forces That Will Shape Our Future*, describes a future when AI will function much like electricity does today, distributed and ubiquitous: "This common utility will serve you as much IQ as you want but no more than you need. Like all utilities, AI will be supremely boring, even as it transforms the Internet, the global economy, and civilization. It will enliven inert objects, much as electricity did more than a century ago. Everything that we formerly electrified we will now cognitize."

Heck, eventually our kids won't be using adjectives like "smart" or "connected" to describe the physical objects in their environment. IoT will be the water they swim in. Everything we make will have predictive maintenance, improved efficiency, better safety, better usability. And everything will be made to order. We won't be cranking out millions of identical widgets and stacking them up in pallets in overseas factories in order to be shipped around the world anymore. We'll be customizing objects much closer to home. As Olivier Scalabre points out, we'll be replacing the classic "East to West" trade flows with regional trade flows—East for East, West for West. "When you think about it, the old model was pretty much insane," Scalabre says. "Piling up stock. Making products travel around the world. The new model—producing right next to the consumer market—will be much better for our environment. In mature economies, productivity will be back home, creating more employment, more productivity, and more growth."

One aspect that I love about all this IoT material is that it's the big, established companies that are rolling out all these new innovations. Schneider Electric (founded in 1836) tracks and monitors elevator usage patterns so that elevators can "default" to busier floors, saving people waiting time. They're also tracking wear and tear, so they can schedule maintenance sessions during low traffic periods. Heidelberg Druckmaschinen (founded in 1850) remotely monitors more than 25,000 high-end printing presses with the help of PTC's ThingWorx IoT platform. Symmons Industries (founded in 1939) builds smart shower systems for hotels that let management teams monitor usage data like temperature, duration, and water volume in order to optimize their utility bills. Gerber Technology (founded in 1968), which makes the textile industrial equipment that creates many of the clothes we wear, used the information generated by those machines to create a new service called YuniquePLM that lets creative fashion teams get the right outfits to the right market at the right time. And that "smart floor" company I talked about earlier (FloorInMotion)? It was launched by Tarkett, a 130-year-old flooring manufacturer from France.

But we're still going to need some new rules. Let's go back to my previous analogy, about the old man waking up in the young man's body. When the seventy-five-year-old man wakes up as an eighteen-year-old, he doesn't get to act like it's 1960 again! In order to fully capitalize on that IoT, manufacturers are going to have to make some fundamental changes in the way they conduct their business. Here's what McKinsey says in one of its recent reports on the topic:

The Internet of Things will enable—and in some cases force—new business models. For example, with the ability to monitor machines that are in use at customer sites, makers of industrial equipment can shift from selling capital goods to selling their products as services. Sensor data will tell the manufacturer how much the machinery is used, enabling the manufacturer to charge by usage. Service and maintenance could be bundled into the hourly rate, or all services could be provided under an annual contract. Performance from the machinery can inform the design of new models and help the manufacturer cross-sell additional products and services. This "as-a-service" approach can give the supplier a more intimate tie with customers that competitors would find difficult to disrupt.

If there's one thing I've learned from working with all these large manufacturing companies, it's that this shift can truly drive growth. What happened to the technology sector is going to happen to the manufacturing sector—I'm sure of it. Why? Because IoT allows you to rediscover your customers. It lets you learn what they really want. In fact, I would argue that the only true competitive advantage is your relationship with and knowledge of your customers. Think about it—what's the first thing your competitor does when you put out a new product? It buys that product on the open market and sends it to the R&D lab, which then proceeds to dismantle it, benchmark it, and reverse-engineer it in a thousand different ways. Your competitors can't do that with the collective intelligence of your customer base. That's something that you, and only you, can own. It's an incredibly powerful advantage.

IoT is about to change the world. But in order to be truly successful at it, we're going to have to rediscover the people who are buying the things that we make.

CHAPTER 8

THE END OF OWNERSHIP

Ownership is dead. *Access* is the new imperative. International Data Corporation (IDC) predicts that by 2020, 50 percent of the world's largest enterprises will see the majority of their business depend on their ability to create digitally enhanced products, services, and experiences. Focusing on services over products is also a sound business strategy. Zuora's *Subscription Economy Index,* which you'll find at the end of this book, shows that subscription-based companies are growing eight times faster than the S&P 500 and five times faster than US retail sales. Our chief data scientist, Carl Gold, put this report together using anonymized, aggregated, system-generated activity on our platform. I urge you to read it— it's a fascinating document, based on billions of dollars of revenue and millions of financial transactions, that has all sorts of industry benchmarks and insights. So far we've seen how subscriptions are changing retail, media, transportation, and manufacturing, but if there's one thing that I've learned, it's that this model is industry agnostic! It cuts through everything. Here are some other verticals in the midst of this shift.

Health Care. When a pharmacy chain (CVS) buys a health insurance giant (Aetna) for $69 billion because it's afraid of an online retailer (Amazon), you know you're witnessing an industry in transition. CVS says the deal will allow it to start offering primary care services at its pharmacies, which makes a lot of sense, particularly when you consider the dispiriting slog that is the typical American health care experience: You visit a doctor, who might send you to a lab for tests, or a hospital to visit a specialist, or a pharmacy to pick up some medication, which may send you back to the doctor if there's a question about the prescription, and so on. There are insurance forms to fill out at every stop; a few months later you're sent a bunch of indecipherable medical bills, and information is scattered everywhere. Is it any wonder the world's most famously customer-focused company sees some opportunities here?

The walls are coming down. This is a $3 trillion industry that is rapidly unbundling. Right now hundreds of new digital services are helping doctors work smarter and letting us detect and prevent health issues so we can stay out of hospitals and retain our independence as we grow older. We're all starting to carry our doctors' offices on our wrists. Fitness trackers are turning into medical wearables that help us diagnose and detect health issues, alert first responders, and even administer medication properly (there are now pills with embedded chips that send out signals on contact with stomach acid). New subscription-based primary care groups like One Medical offer same-day appointments and Apple store–like customer service, and holistic providers like Magellan Health are connecting behavioral, physical, pharmaceutical, and social needs. All of these new channels are transferring autonomy and agency from hospitals to patients, which is a good thing.

Government. Here in the United States, it's easy to grumble about government inefficiency when it comes to basic services: paying taxes, registering a business, getting a driver's license, paying a toll. Everyone understands the pain points. In other places, though, all these points seem to be associated with considerably less pain! Estonians don't just pay their taxes online, they simply authorize an online tax statement that's been fed real-time financial data over the course of the previous year. One-click taxes. Rwandans apply and pay for driver's licenses over their phones. In Sweden, a digital service allows medically trained citizen volunteers to be instantly alerted if there is a heart attack victim within five hundred meters of their location. With just one citizen ID, New South Wales residents in Australia can log on to Service NSW for more than eight hundred different kinds of government transactions, or visit one of more than a hundred dedicated offices with concierges and free Wi-Fi, which they call "one-stop shops." Thanks to digital services, governments are moving beyond fixing potholes and collecting library fees toward becoming platforms for civic creativity.

Government spending represents significant percentages of national GDPs all around the world—here in the United States it's roughly a third. That's a lot of locked-in value. Think about how much inefficiency could be solved with secure citizen IDs cutting across all those agencies. In the United States we're particularly sensitive to privacy issues, but the choice between signing my life away to Big Brother and having my information scattered across dozens of hopelessly siloed government databases is a false one. Right now, virtually every single one of my interactions with my state and federal government is a depressing time suck. It's not just consumers who are demanding more transparency and automation—it's citizens as well. Fortunately, governments are starting to listen.

Education. Many of us are working in positions that didn't exist a few years ago. In professional environments that are constantly in flux, continuous learning is an imperative. And yet higher education is the only industry in the world that fires all its clients after four years. How crazy is that? Take business school—much of today's MBA experience could be summarized as two years of friendly networking coupled with some ostensibly useful coursework. But what happens when an MBA who is ten years into the workforce gets promoted and suddenly needs to brush up on a new skill set or learn subject matter that was missing entirely from their old coursework (like most of the material in this book!)? Well, usually this results in lots of frantic Googling.

Instead, as a couple of Wharton professors have suggested, what if business school consisted of ten months of campus coursework, followed by a lifetime of personalized, instantly accessible online courses featuring the latest available research? What if school never ended? Today many colleges and universities are experimenting with MOOCs (massive open online courses), but I think the concept needs to be extended well past graduation day. We've all seen the explosive growth of professional learning platforms like Lynda.com (now part of LinkedIn), Kaplan, Udemy, and dozens of online coding academies. Heck, I recently saw a bunch of eight-year-olds taking a class at an Apple store in New York. And in the textbook industry, newcomers like Chegg as well as established publishers like Houghton Mifflin Harcourt are offering online rentals and interactive content to help students save money and learn more effectively.

Insurance. Today most of your premiums are determined by all sorts of actuarial factors outside your control. But what if you're someone who exercises a lot? Or if you don't drive all that much? What if you don't drink? Shouldn't your premiums be reflective of that? Health IQ, for example, gives active people lower health insurance rates with the help of a clever health quiz. And in an era when more and more of us choose to live in places part time, work in places part time, and subscribe to things rather than buy them, fixed long-term insurance contracts built around static assets just don't make sense anymore. Why can't I get liability insurance for part-time work?

Sixty-five percent of drivers overpay on insurance premiums in order to subsidize high-mileage drivers—Metromile fixes that by offering people "pay-per-mile insurance" as recorded by a simple connected device that fits in your car's OBD II port. Lemonade, a new home and rental insurer start-up, is trying to avoid the industry's inherent conflicts of interest (i.e., insurers drive profitability by denying claims, people are incentivized to embellish claims) by taking a flat, fixed monthly fee, paying off a few expenses, and using the rest to pay out claims. Insurance companies are finally taking advantage of what Deloitte calls "flexible consumption" models in order to pay attention to customers, not just cohorts.

Pet Care. This is a $100 billion global industry that's growing much faster than many other consumer packaged goods verticals, but it suffers from the same systemic pain points, including reseller channels that sever direct customer relationships. Digital services are changing all that. Retail pet food companies are turning into digital pet health services. Today you can go to My Royal Canin, fill out a quick profile, and based on your pet's age and breed they'll send you the right kind of food; relevant suggestions around nutrition, health, and grooming issues; and access to a dedicated staff of veterinary technicians. What about when they get sick? Budgeting for your own health care is difficult enough. Trupanion offers health care insurance that covers 90 percent of a veterinary invoice for life.

It's not surprising that there are lots of parallels here with the market for parents of young kids—according to business solutions agency Gale, 44 percent of millennials see their pets as "starter children." Every cool new service advertised on your favorite podcast has a pet equivalent. There are collar tracking services, monthly box services like BarkBox, wellness monitoring services, location-based apps that help you locate pet-friendly parks and establishments, connected food dispensers, online breeding and training courses.

Utilities. The big electricity, gas, and water companies used to follow a pretty standard playbook: spend lots to build a giant utility plant, spend lots to transfer that utility across a very long distance, then charge lots in order to pay down your infrastructure debt. Today things are changing. As *The Economist* notes, "Not only are renewables playing a far bigger role; thanks to new technology, demand can also be tweaked to match supply, not the other way round." New consumption-based digital services like SolarCity are enabling solar-powered homes to sell electricity back into the grid, for example, and home energy management systems like Nest are helping utilities lower their power consumption at peak times in order to avoid shortages. Lots of people aren't getting electricity bills anymore—they're getting electricity checks.

Yes, utilities are some of the oldest subscription-based businesses around, but today they're moving away from big, monodirectional channels to smaller, responsive networks (notice a recurring theme here?). Thanks to sensors and connectivity, the entire industry is starting to "wake up" and unlock all sorts of new value: French electricity provider ENGIE (which traces its roots to the company that built the Suez Canal) now has an app-based service that lets you book home service appointments; Schneider Electric (founded in 1836) partners with municipal districts not to sell them more power, but to help them *reduce* energy consumption by 35 percent. A new solar start-up in Brooklyn called LO3 Energy is using blockchain technology that lets you sell your solar energy to your neighbors. All these new services aren't just about adding convenience, they're about accelerating outcomes.

Real Estate. For generations, we've been taught that buying and owning a house is an expected and necessary part of becoming an adult. Clearly, that's no longer the case. Lots of younger people disagree with that idea, and for very good reason. They want to have more flexibility, more choices, and the ability to mix up their environments when they feel like it. In our office at Zuora, we give everyone a desk, but lots of people can't stand them. They're much more comfortable taking their laptops to a sofa in a lounge space, putting on their earphones, and cranking away. Companies like WeWork and Servcorp have figured out that they can make more money per square foot catering to these new preferences. Companies aren't as interested in onerous long-term property leases anymore—they want the flexibility to expand or decrease their footprint as they see fit.

More and more of the physical world is becoming unlocked. Mobile workers and entrepreneurs are ditching coffee shops for shared workspaces. Every day, more than 1.2 million people go to an "office" that's full of other freelancers or small corporate teams. People are finding all sorts of cool new getaway experiences on vacation rental sites like VRBO or digital nomad platforms like Roam. In response to Airbnb, hotel companies are realizing that they're in the business of creating compelling travel experiences, not just putting their names on big resort properties, so they're diversifying into apartment rental platforms. Subscription-based digital services are a big part of the business models powering these sites, whether you're signing up directly with HomeAway or your real estate agent is taking advantage of a professional service on Zillow to reach more buyers.

Finance. For way too long, the banking industry has treated the internet like just another point of customer contact, a virtual teller station. Whether you're a small business or an individual, you're free to look up your account and swap some money around online, but the basic mechanics of accepting deposits, raising capital, making loans, or buying securities have stayed firmly inside the vaults. According to *HBR*, a recent analysis by Bain and SAP found that only 7 percent of bank credit products could be handled digitally from end to end. That's all about to change. New fintech services are dispersing those mechanics in all sorts of interesting ways.

If the health care industry is moving to your wrist, the financial sector is moving to your phone. Wealthfront uses algorithms to help you invest responsibly, save for retirement, or contribute to their own 529 college savings plan. Robinhood uses a subscription model to help investors stop paying ten bucks a trade. Adyen helps companies do business all around the world by processing almost every payment method under the sun. Venmo is a digital wallet that helps you split the bill or cab fare with friends (if you've never heard of Venmo, ask your younger colleagues about it). Everything is up for grabs, including the whole concept of fiat currency.

A NEW PATH TO GROWTH

There's lots more we could talk about. We work with companies in agriculture, communications, travel, wellness, telecommunications, life sciences, aviation, food and beverage, fitness, gaming. But here's the through-line: subscriptions lead to growth. Once customers can get the outcome they want, without having to worry about owning the physical assets, that's where the demand goes, and that's where new revenue streams are created. Every sector on the planet has the same potential to latch on to the same kind of growth that the technology industry has recently enjoyed. As Andy Main, the head of Deloitte Digital, said at our *Subscribed* conference in San Francisco:

It's an open playing field today. And that open playing field has re-sulted in brand-new experiences, different models. And that means you need to wire up your business in different ways. It's all about competing on experience, and thinking about your value proposi-tion, and how you orchestrate your business in order to bring that value to life. Experience is the new frontier of competition.

This is where everything is going. All of these case studies, ar-ticles, and studies point to a world where you and I can tap into services for everything we need, a world of freedom. But what will it take to succeed in this new open playing field? How can you set up your organization for success in the Subscription Economy? Well, it may involve having to negotiate your way through a WTF moment. Let me explain.

PART 2

SUCCEEDING IN THE NEW
SUBSCRIPTION ECONOMY

CHAPTER 9

THAT WTF MOMENT

If digital transformation means roles are changing, then by definition the way companies function as a whole is changing as well. To help us visualize how much transformation is needed to become a subscription business, let's run through a thought experiment involving a fascinating industry that we haven't talked about yet—video games.

Games are arguably bigger than Hollywood, and there are several parallels between the two industries. Much like Hollywood blockbuster franchises, every big game brand is a franchise, and their revenues are massively front-loaded in boom-or-bust release weekends. A major game may cost up to $60 million to develop, and sometimes twice that number to market. In much the way you'd rather produce *Titanic* than *Gigli*, you'd rather release *Grand Theft Auto V* (which made more than $800 million in twenty-four hours) than *Transformers: Rise of the Dark Spark* (which did not). Game studios typically spend two years working on a title, blast it out to as many sales channels as possible on its launch date (mostly stores and consoles), and then hope that the customers are waiting for them at the end of those channels.

Video games are also following the same general media consumption trends as movies—retail disc revenue is plummeting, while online revenue from streaming and subscriptions is on the ascent. And much like Hollywood, the video game industry is keenly interested in pursuing a multiscreen strategy: a console, a phone, a device like a Nintendo Switch, a store, a streaming site like Twitch, or even Madison Square Garden or the Staples Center for a big e-sports competition. Your average video game player is having all kinds of experiences with video games, generated by a myriad number of channels. They're playing online with friends, paying for downloadable content (DLC), conducting microtransactions, playing a mobile version on their phone, watching really great players on Twitch (and maybe sponsoring those players), and going to conventions.

Now let's say you're a developer with a big franchise game called *Starship Blasters*. Every two years you come out with a bigger, better *Starship Blasters* game, with new characters, crazy new adventures, and (of course) better blasters. But these sequels are getting more expensive to make, and they're making less money with every new release. You know that in two years, probably only half the people who bought *Starship Blasters III: Still Blastin'* will buy *Starship Blasters IV: Blastageddon*. And you know the kinds of experiences your player wants to have with their games just don't align to a biannual release schedule.

So, you figure, if you can spread the $60 game cost over a year for $5 a month, and hold on to that player with lots of cool new downloadable content, you'll do better in the long run. They won't get bummed out paying lots of money for the occasional dud game, and your company will benefit from a stable revenue model that's less reliant on the ups and downs of Hollywood economics. You'll also start every quarter with a platform of recurring revenue to invest as you see fit. As *Final Fantasy* producer Naoki Yoshida told *GameSpot*:

> *With the subscription model, you have that constant flow of revenue. As game developers, creators of games, we want to be able to continue providing the best gameplay experience and sustain that. Of course, the initial subscriber numbers might not be as many as the free-to-play model, but we have that constant stream. We're not thinking just about the business of the moment. We want to think about the long term and being able to have the funding to continue making updates. Some people might be focused on quickly gaining revenue, but you have to think about the long term.*

So it's decided. No more big splashy releases every two years. For just five bucks a month, now your player gets *Starship Blasters* as a service—constant innovation, rolling updates, continuous engagement. Everybody wins. You slave away on an impressive PowerPoint presentation, the big board meeting happens, and then the company-wide email goes out, and every department receives a clear set of directives and deliverables around shifting to this new business model, with a final launch date.

And the response from your company? A big, massive WTF.

Marketing is irate because they don't have a big blitz launch day. Development is having kittens because they've lost their entire production schedule. IT is highly indignant because the millions of dollars they just invested in a new set of business systems just became instantly obsolete. And finance is not exactly overjoyed at the imminent prospect of seeing their quarterly revenue numbers tank.

So how do you do it? How do you make the shift? That's what

the second half of this book is all about—building a subscription culture. Let's go back to this diagram:

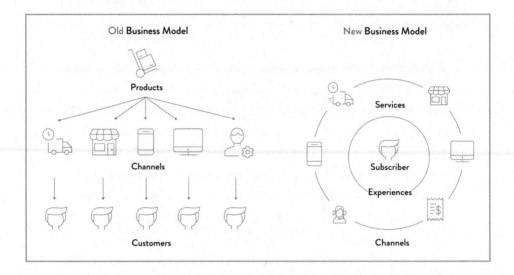

There's your video game player on the right. They're not hiding behind your channels anymore. But how will this shift affect the way you run your company? None of this stuff is easy. The imperative to make this shift is loud and clear, but how can you help each one of your employees make the leap? How will you respond to your own company's WTF moment?

Let's start with your product managers, engineers, manufacturers, designers. They have all kinds of positions, but ultimately they're responsible for product innovation in your company. In the old model their job was to do the market research, run some focus groups, draw something up on graph paper, figure out how to manufacture it at the lowest cost, get it put together, and then put it into the marketplace. If you sold lots of units and exceeded your projections, then you had a hit product on your hands and made a lot of money. If you didn't, then you didn't. But what happens when a static product turns into a living, breathing experience? What happens when the customer comes in out of the cold and sets up shop squarely in the center of your development process? How do you happily surprise that customer on a consistent basis? Based on all this new usage behavior you're seeing, how do you build better pathways where you see that people are walking?

What about the finance team? The controllers, the CFOs, the operation folks? The job used to be about tracking expenses and tying them back to that almighty unit of sale: How can we improve our marginal cost? How much money do we need to give to our distribution channels? How much overhead do we have? In the case of *Starship Blasters*, your finance team is focused first and foremost on unit sales, with digital services as an important, growing, but ultimately secondary source of revenue (a lot like a certain company based in Cupertino, California). Well, things are no longer so simple.

Today finance teams are experimenting with a whole new rulebook and set of metrics: costs of customer acquisition, lifetime customer value, annual recurring revenue, average revenue per customer. The five-bucks-a-month plan sounds like a great idea, as long as your players hang on to the game for at least a year. What about the ones who drop off after a couple of months? How are they accounted for in your business plan? Today finance teams are increasingly defining the operating metrics that are distributed throughout entire organizations, such as pricing, packaging, and analytics. For example, how does Netflix justify spending $8 billion a year on shows that it doesn't really "sell"? Its finance team obviously knows something that we don't.

How about the CIO and IT group? How is their job changing? Over the last twenty years, the focus has been on efficiency through standardization: How do I lower my inventory? How do I shorten my supply chains? How do I get a product from point A to point B faster and cheaper, thereby lowering my per unit cost, in order to give my company a competitive advantage? How do I maintain a consistent system of record? Everyone installed fancy ERP systems to help manage their supply chains. The goal was standardization across systems of record. If there was a little extra in the budget to try out some new technology, then great. But for the most part IT folks were down in the engine room, shoveling coal and keeping the lights on. But what happens when your IT infrastructure is built around customers (or players), not units of sale (or discs of *Starship Blasters*)?

Today the analyst firm Gartner says that "IT is moving away from focusing on systems of records to focusing on systems of innovation." What does that mean? The product folks are going to IT and asking, "How do we launch new services and iterate?" The sales and marketing folks are going to IT and asking, "We have all of these pricing and packaging ideas, but how do we try them out? And how do we do that quickly?" The finance group is banging on the engine room door: "We need a completely different view of the business, beyond what our traditional finance systems are able to show. How can you help us?" How does your IT group respond to these new kinds of challenges?

Finally, how does this model affect your sales and marketing efforts? Look at it this way—what's the difference between selling a transaction and selling a relationship? As Anne Janzer notes in her great book *Subscription Marketing Strategies for Nurturing Customers in a World of Churn*, "Marketing is no longer just about getting to the sale. To keep subscription customers renewing and re-engaging, you have to provide real value and solve problems." In the case of *Starship Blasters*, that imperative puts a premium on making a great game that's constantly evolving, so maybe you redistribute some of your billboard advertising budget toward hiring more developers and throwing more conventions. Maybe. More traditional sales teams are faced with a related challenge: today's prospects already know lots about your company, because there's so much information on hand, but many times this means that they're more confused than ever. They might not even know the right questions to ask. How do you start that conversation? How do you sell to someone when you know you're going to be on their contacts list for the foreseeable future?

TEARING DOWN THE SILOS

The twentieth-century organization resembled a sequence of stovepipes—various business departments that generally kept to themselves. And in the past, these strictly siloed functions made sense. Marketing did the research and passed it off to the product group, which took the specs and made it happen. They passed it off to the sales team, which pitched it like crazy. Finance counted the beans. IT got called when someone forgot their password.

This way of working—organizing your company around a specific product line—made sense when everything was about scale, consistency, and any color as long as it's black. It let management have a direct line of sight into development cycles and sales performance and provided for clear team ownership accountability. But while this product-based approach had some benefits in terms of assessing core transactional issues, it came at a tremendous cost. It created a siloed organization that lacked any kind of coordinated vision. It fostered internal competition for customers. And it contributed to a myopic mindset that blindsided management to sudden changes in the marketplace. These companies were built backward—oriented around the products, instead of their customers.

But unfortunately, a lot of the time this structure also *worked*. It didn't work particularly well when it came to consumers, but it shipped units and kept boardrooms happy. It was a functional but mediocre way of conducting business. And there is nothing more dangerous than profitable mediocrity. The golden era of postwar corporate dominance, when this original product-based management structure was created, was enabled by a relatively passive consumer base.

This is clearly not the case anymore. In this new world, with the customer at the center, these silos must come down. How can you create a new subscriber experience that delivers outcomes if everyone has their heads down? How can you deliver business model innovation, make the right choices across how you innovate, how you go to market, how you approach your fundamental business model, if teams are separated by walls? Let's take a closer look at the new rules for subscription businesses, starting with the heart of your organization—the designers and inventors who are now tasked with turning a great product into a great service.

CHAPTER 10

INNOVATION: STAYING IN BETA FOREVER

A funny thing happens to software companies right after they launch their first SaaS offering. It's the same thing that happens to media companies when they ask their users to register for the first time, or retailers that start tracking purchases back to specific customers, or industrial manufacturing companies after they put sensors in their equipment. All of a sudden, they can see what their customers are doing! It's a really incredible experience watching the dashboard light up for the first time. I remember the first time we experienced it at Salesforce. We instantly became hungry for more information, and it profoundly changed the way we made decisions, allocated resources, and built new services. It changed everything.

GMAIL AND THE NEVER-ENDING PRODUCT

I have a personal theory that lots of people got introduced to a new kind of product development philosophy when Gmail arrived. When it first launched on April 1, 2004, Gmail had the word "BETA" stamped on its logo. Millions of people signed up for it, but it was still a beta product. In fact, it was a beta product for five years—it didn't finish its "testing" phase until July 7, 2009. So what took them so long? Why did they eventually decide to drop the beta moniker? Believe me, it didn't have anything to do with the engineering team calling it "finished." The real reason was that big Fortune 500 companies that were used to buying Lotus Notes or Microsoft Exchange wanted to buy Gmail for their companies, but their procurement offices wouldn't let them buy a beta product! Google's response? Just update the logo.

Here's Google's blog post announcing the end of "beta status" for Google Apps on July 7, 2009 (the last sentence is my favorite):

We're often asked why so many Google applications seem to be perpetually in beta. For example, Gmail has worn the beta tag more than five years. We realize this situation puzzles some people, particularly those who subscribe to the traditional definition of "beta" software as not being yet ready for prime time.

Ever since we launched the Google Apps suite for businesses two years ago, it's had a service level agreement, 24/7 support, and has met or exceeded all the other standards of non-beta software. More than 1.75 million companies around the world run their business on Google Apps, including Google. We've come to appreciate that the beta tag just doesn't fit for large enterprises that aren't keen to run their business on software that sounds like it's still in the trial phase. So we've focused our efforts on reaching our high bar for taking products out of beta, and all the applications in the Apps suite have now met that mark. . . .

One more thing—for those who still like the look of "beta," we've made it easy to re-enable the beta label for Gmail from the Labs tab under Settings.

Hilarious—if you miss the old "beta" logo, you can still find it in your settings! The point being: this four-letter word means absolutely nothing to us.

I think this is a really important document—it spilled a fundamental principle of modern software design out into the world. Gmail marked the end of seasonal boom-and-bust product cycles and the birth of the *never-ending product.* What do I mean by that? Well, the original point of beta was to let you put a product out before it's ready, gather up lots of feedback from your customers, and then incorporate that input before you finally freeze the product and ship it. After all, a key aspect of "agile" product development is the participation of customers and key stakeholders in the development process before the final product goes live in order to anticipate outlier scenarios and generally help out with QA (quality assurance) efforts. The Gmail team took this idea a step further—they decided to ignore the last part. What they realized is that if you can get out of that mindset and think about what you're making as a living, breathing experience instead of a static product, then why not enlist your customers as innovation partners all the time? Why not stay perpetually in that beta mindset?

The Manifesto for Agile Software Development was put together by a group of developers at a ski resort in Utah in 2001. It contains four simple but powerful value comparisons: individuals and interactions over processes and tools, working software over comprehensive documentation, customer collaboration over contract negotiation, and responding to change over following a plan. You can apply these principles to any kind of subscription service. Innovation doesn't happen in a vacuum. It's the result of iterating a concept over a period of time. Big "boom or bust" product launches can actually be a recipe for burnout: they result in unhealthy peaks and troughs of productivity and inspiration. The idea is to create an environment that supports sustainable development—the team should be able to maintain a constant pace of innovation *indefinitely*. That's the only way to stay responsive, to stay agile.

And now that concept—always be listening, always be iterating—is spreading out into every industry on the planet.

KANYE WEST AND THE FIRST SAAS ALBUM

Take Kanye West's album *The Life of Pablo*, which I mentioned earlier. He dropped it on Tidal on February 14, 2016, and to nobody's surprise, everyone freaked out and started streaming it like crazy. But then something weird happened . . . he kept working on it! He was still adding vocals, changing around lyrics, and tweaking the song order weeks after it officially released. When musicians used to ship an album, there was a finality to it: the reviews came in, the fans either loved it or hated it, and that was it. But streaming put an end to that.

Now, I admit that you can get into trouble with this concept (e.g., George Lucas messing up the original *Star Wars* trilogy with all sorts of stupid special effects), but why can't you go back and listen to the same album a year from now, two years from now, and have a slightly different experience? Talking to Atlanta radio station V-103, Kanye explained that the cover of his previous record, *Yeezus*, was just a plain shrink-wrapped CD because it was "almost like the death of the CD, like an open casket. It was like 'Look at this CD, that we've been looking at our whole lives, just look at this for the last time, because you ain't gonna be seeing it too much more.'" Kanye put out the first SaaS album.

Kanye put his listeners, not his album, in the center of his creative process. He shortened his product development cycle through experimentation, validated learning, and iteration. And he helped to create a virtuous feedback loop whereby customer feedback helped inform product development. By putting it out there—and asking subscribers to support it—Kanye successfully fed his sales funnel without having to wait for a "final" product. Instead, he gave himself the space and resources to tinker with his music, optimizing it as part of an ongoing deployment cycle. Although I'm pretty sure he wouldn't have put it this way.

GRAZE: THE AGILE FACTORY

To see how powerful this whole concept of continuous innovation can be, and how it's not just limited to software or digital media, I want to share a story about a UK snack box company called Graze. They're like a Pandora for snacks. Every couple of weeks they send me a box of four different kinds of treats, and I give them feedback in a simple online form: "I liked this. I didn't like this one: don't give me anything like that again. I may be able to check out some more options around this one. Keep the popcorn coming, etc." They've got that suggestion engine algorithm that you're seeing everywhere these days. Pretty cool stuff.

When you learn more about the company, though, it gets even cooler. I talked about agile software development—Graze has an agile *factory*. At our *Subscribed* conference in London, Graze's CEO, Anthony Fletcher, pulled out his phone and said: "I can run my factory from my phone: the supplies, the distributors, the packaging. Every box that I ship is intended for one person and one person alone." That's pretty incredible, but that's not where the story ends—here's my favorite part.

Graze recently launched in the United States, where traditional English snacks like marmite and salt and vinegar crisps don't exactly fly off the shelves. Its CEO said, "I used to work for an energy drink company, a CPG company. When we went into overseas, we would spend millions of dollars on market research to understand the tastes and the preferences of that new market. Still, more than half of the time, we would get it wrong. We would dump something in the marketplace, and if it was a miss we would have to regroup and figure out how to launch again in a year. At Graze we didn't spend a dime on market research when we launched in the US. We just took our existing product line and dumped it on the US market, because the system adjusts itself." The Graze team just sat themselves in front of the dashboard and waited—after a few days the spicy barbecue flavors started rising to the top, while the chutneys took a nosedive.

No more focus groups, phone surveys, or user interviews. Also, no more hoping and praying you get a hit product. Why? *Because the market research is already baked into the service*. The Graze team had their US distribution completely dialed in three or four months because they could see what their customers were doing. They could drop that factory anywhere in the world and it would immediately start listening, learning, and optimizing itself. When you design your service in conjunction with your subscribers, and inform that service with usage and behavioral data, you can make something that they really love and that evolves with their needs. The Gmail team realized this, and so did the Graze team.

NETFLIX: NO MORE PILOTS

Pilots are a fundamental part of the network television development process. Rather than shell out for a whole season of a TV show that may or may not work, big TV studios pay to shoot trial episodes, then show them to test audiences in places like Las Vegas (development people love Las Vegas because the visitors there represent a pretty accurate cross-section of America) in order to gauge their reaction. TV shows are expensive to produce, so pilots are a way for television studios to hedge their bets. A typical pilot season is a brutal, *Hunger Games*–like scenario: several hundred pitches get culled down to several dozen scripts that in turn make it to fifteen or twenty actual pilot episodes. Pilots have short, difficult lives—*Variety* estimates that less than a quarter of them eventually become full television shows. They are market research.

Netflix doesn't use pilots. Never has, never will. Now, make no mistake—Netflix has had its share of duds, but along with HBO it has been extraordinarily successful in creating those zeitgeist shows that everyone talks about: *The Crown, House of Cards, Orange Is the New Black, Stranger Things*. With *House of Cards*, its first foray into original content, Netflix knew that the little-known British version had done notably well on its platform, and it also knew that David Fincher, Kevin Spacey, and Robin Wright were all very popular with its audience. Netflix knew there was audience enthusiasm for political drama, so a very promising Venn diagram was starting to take shape. But it also knew that this was going to be a dense, multilayered story that would not be well served by a pilot episode. So it wrote a check for the whole show. Jonathan Friedland, the company's chief communications officer, told *The New York Times* that "because we have a direct relationship with consumers, we know what people like to watch and that helps us understand how big the interest is going to be for a given show. It gave us some confidence that we could find an audience for a show like *House of Cards*."

We all know how committed Netflix is to user data. It looks at millions of customer touch points (or "plays") a day, including when you pause, rewind, and fast forward, as well as user ratings, searches, geolocation data, viewing times, device information, and social media feedback. If you play one title, then what did you play before, or after? What did you quit watching after five minutes? It also tags every single show on its platform with more than a hundred different designations, including things like violence levels, geographic settings, what time of year the story is occurring, and even what kinds of jobs the characters have. "It is a beautiful thing being a subscription service," Todd Yellin, Netflix vice president of product, told *The Guardian*. "We have nothing to do with advertising, it becomes less about ratings. The days of pure popularity as a yardstick of success are over. It leaves the individual quirks and quirks of people's taste in the dust. We share all the data with the Los Angeles programming team, to see how it compares to the shows they are thinking about. User data helps us decide to initially buy the show and to renew it for another season. Traditional networks and cable networks don't know that stuff."

No need for test audiences and ratings cards. Netflix is still competing in a creative industry, where there are successes and failures, but it has a "giant brain" that is conspicuously lacking in network TV. With subscription services, all the insights you need are sitting right there in your system.

STARBUCKS AND SUBSCRIBER IDS

Early in 2017, Starbucks got a lot of press because it had a fairly unique problem—its mobile app was becoming *too popular*. Too many people were ordering their coffee drinks ahead of time for pickup, leading to long lines at the stores. "We are now laser-focused on fixing this problem, but the nature of it—too much demand—is an operational challenge we have solved before and I can assure you we will solve again," ex-CEO Howard Schultz said on a Starbucks earnings call. Starbucks eventually fixed the issue by dedicating baristas and pickup areas during peak hours.

That story isn't really about how mobile technology is rapidly changing our daily retail transactions (well, maybe it is a little bit). But the bigger story, which was mostly missed by the business and tech media, is about the power of the Starbucks ID. Once you establish a secure identity with a customer that includes things like purchase activity, payment information, perhaps some demographic details, or maybe some location alerts, then you can do amazing things. Today, more than 13 million people are enrolled in Starbucks' rewards program, which now represents more than a third of US company-operated sales. One in every ten transactions in a US Starbucks store is handled by its mobile app, which tells you when your order will be ready and how long it will take you to arrive at the closest store. Starbucks is working toward getting rid of lines entirely. But it all starts with that ID.

And here's a scenario likely to greet regular Starbucks customers pretty soon—say you're in another city on a work trip, but you still want to get your latte fix. What happens when you drop into the Starbucks outside your hotel? "We detect you're a loyal customer and you buy about the same thing every day, at about the same time," Starbucks CTO Gerri Martin-Flickinger recently told *CIO*. "So as you pull up to the order screen, we show you your order, and the barista welcomes you by name. We also show you your favorite treat in a picture at the same time. Does that sound crazy? No, actually, not really. In the coming months and years you will see us continue to deliver on a basic aspiration: to deliver technology that enhances the human connection."

Starbucks is clearly taking a cue from the big GAFA companies: Google, Amazon, Facebook, Apple. In China, their equivalents are known as BAT: Baidu, Alibaba, Tencent. But what's the common theme among all of them? Subscriber IDs. They all have dashboards that let them see what their customers are doing, so they can make smarter decisions around where to allocate resources and which new services to spin up—in Starbucks' case, maybe it's how many drink reward "stars" to give out, or where to open a new store. The Gmail team has the dashboard. Netflix has it. Here's an interesting question—who *doesn't* have it? What about the rest of your daily transactions? What about the other companies you deal with on a regular basis? Today, you probably don't walk around with a Coke ID, a Nike ID, a L'Oréal ID. But if you're a fan of those companies, I predict that pretty soon, you will.

CHAPTER 11

MARKETING: RETHINKING THE FOUR P'S

Quick—when I say marketing, what do you think of? I bet you visualize Super Bowl ads, or Don Draper from *Mad Men*. Maybe it's the Apple 1984 commercial, or the Geico gecko. Or all those crazy dotcom commercials from 2000: the Pets.com sock puppet, the Webvan fleet. Loaded with VC cash, the marketing departments of all those companies decided to take all that money and do what they do best: advertise. Because that's what you did if you were in marketing. You advertised.

Back in the old days of doing business, there was a very good reason marketing departments thought this way. Let's take another look at the left side of the diagram:

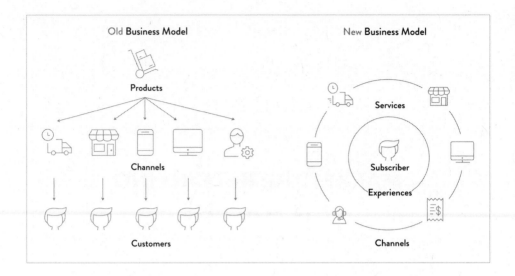

Remember, the goal in this world was to sell product, and lots of it. It was an "asset transfer model," and that asset was transferred through distribution channels. The marketing department's job was to focus on "push" and "pull" techniques. "Push" meant driving the product through the channels, so you gave money and resources to the channels in order to promote your product over the competitors': channel rebates, preferred in-store placements, commissions to salespeople. But you also focused on "pull," which was about driving customers to the channels to ask for your product. And how did you do that?

Well, you conveyed exceedingly pleasant images of what the world would be like for someone if they were to own the asset in question! And so we got the postwar ascendance of advertising: the Marlboro Man, the world buying a Coke sing-along. Marketers and advertisers prided themselves on their Clio awards, Super Bowl spots, billboard one-liners. The product was really just a commodity—the fun, creative part was selling it. And if it didn't exactly work out for the buyer, well, there would always be more products to pitch.

Today, however, things sure feel different. All of a sudden we're surrounded by all these big, successful companies that don't appear to spend a dime on advertising—traditional advertising, anyway. When was the last time you saw a Netflix red envelope in a newspaper? I think everyone would agree that brands are still very important, but today you communicate your brand through *experiences*, not ads. The best sales pitch for Netflix is binge-watching a great Netflix show. The same principle applies to buying glasses from Warby Parker. Or conducting a Google search. Or looking up a prospect on Salesforce.

At the same time, we're also hearing a lot more about how all these companies have in-house teams of "growth hackers," which on a surface level sounds a lot like, well, marketing. They're trying to come up with smarter ways to drive sales. But these folks tend to reject that label. Stitch Fix has more than ninety data scientists on its payroll. These people aren't thinking of snappier punch lines for billboards; they're looking for ways to optimize growth *within the service itself.* It's almost as if the engineers have taken over the marketing shop: building freemium models, creating upgrade incentives, offering in-app purchases.

So what's going on here? Well, let's go back to the right side of the diagram. In this new world, you have to start with the customer. Actually, this has been a big trend for the last twenty years—the whole concept of one-to-one marketing, emphasizing individuals and personalization. But here's the thing—there is no better demonstration of "one-on-one" marketing than a subscription service because that's exactly what a subscription service is— a one-on-one relationship.

If you're in marketing, you've been trying to collect information about your customers since . . . well, since as long as you can remember. You've been paying big bucks to get demographic information from things like Acxiom, BlueKai, Experian. Heck, last year companies spent more than $10 billion on this stuff in the US alone. But in the Subscription Economy, your engineers and product developers have already taken care of this—they've given every one of your customers a subscriber ID and tracked every transaction and process back to those individual subscriber IDs. What a gold mine! This should be marketing nirvana! There's no more looking elsewhere for answers—they're all right there.

And guess what? This also completely changes the four P's. What are the four P's? In Marketing 101, every MBA gets taught the four P's. The basic idea is that your marketing strategy should focus on four areas of concentration. I'm simplifying, but here's the basic framework:

Product—you should make and package something that people want.

Price—your product should have a competitive price that makes sense for both your company and your customer.

Promotion—your product brand should be advertised through attractive channels (presumably by attractive people).

Place—your product should be distributed and sold in convenient and compelling locations.

But what happens when that first P—Product—gets turned into an S? You can no longer look at the other three P's the same way. This is the story of how everything in marketing changes. Let's work backward and start with your distribution channel.

PLACE

Here's one of the most common questions we get from companies that are shifting to subscription models: What do I do about my channel? Channels are a big part of the world we live in. General Motors relies on car dealerships. Cisco relies on software resellers. Procter & Gamble relies on retailers. Magazines still depend on newsstands. Taylor Swift relies on Spotify (except when she doesn't).

The problem is that today most manufacturers don't own the customer relationship—the channel does. Maybe it's a retailer. Maybe it's a distributor. But when manufacturers start making noise about establishing a direct relationship with the end customer, they can freak out their distributors. So what do you do about it?

If the whole promise of a subscription model is finally letting you build a true 1:1 relationship with your customers, understanding what they are doing and being able to guide them through a subscriber journey, then how do you make that work with your channels?

Well, you look at companies that have successfully managed the transition. Take Autodesk, which makes software for architects, engineers, and designers. They spent several years making the shift. The first thing they did was reinstrument their software for subscriptions: establish subscriber IDs, push updates out on a regular basis. They reimagined their software from a static product based on a traditional waterfall development process to an agile, ongoing service. They switched into beta mode.

Once they taught their core product how to act differently, it was time to educate their resellers. They invested a lot of time and resources into general education about how subscriptions work: workshops, white papers, seminars, etc. Remember, these resellers were used to selling big on-premise contracts. Autodesk wisely decided not to deep-six those big deals immediately. Instead, they offered them something extra—an annual maintenance plan that they could sell as an added service. This accomplished two things: it got their resellers familiar with the idea of services, and it also established an annual relationship cycle, a rhythm.

Then they taught their resellers about managing a relationship over time, as opposed to re-upping a contract once a year. They gave them a practical schedule—the first three months pay attention to adoption, the next six months pay attention to usage, and the final three months start preparing a package for renewal and relevant potential upsells. Also, Autodesk was suddenly swimming in all sorts of new user data as a result of reengineering their software. So what did they do with it? They shared it with their resellers. Every reseller got relevant information about their customers— behavioral data they hadn't seen before.

Autodesk didn't cut out the resellers—in fact, they doubled down on their commitment to them. They used the newfound knowledge of their subscribers to make the channel even more successful, in a way that individual resellers couldn't do themselves. This same dynamic cuts across all verticals. What about the automobile industry? Thanks to services like GM's OnStar, for example, garages are getting automated requests for maintenance, and they can schedule and execute those appointments much more efficiently. What about retail? Someone who's way more invested in their guitar because they've been learning from Fender Play has all the more reason to visit Guitar Center. The Subscription Economy isn't just about win-win models. It's also about win-win-win.

PROMOTION

If the promotional aspect of the marketing mix used to rely on some combination of pull (big splashy advertisements) and push (commissions, marketing development funds, channel rebates), what happens now? Do you just take your promotional budget and hand it over to the engineers? Well, no, not exactly. Brands are still important. But these days, they are increasingly communicated through *experiences*. There's "the sign-up" experience, the "first time you try it out" experience, the "this is cool I'm still using it" experience, or the lack thereof. Lots of people bought Amazon Echos thinking they were going to get to have conversations with a nicer, nonlethal version of HAL 9000, but they wound up with expensive clock radios.

So how do you get people into that experience? Well, thirty years ago, the only way to do that was through ads—the pull. Then people moved to Google to find things, so search became the main driver. Now it's shifted to social—Facebook, Twitter, WeChat, LinkedIn, and a million private networks. So do you just throw all your promotion budget at Facebook? Again, not exactly. Today most of our commercial transactions are mediated through social experiences. Word of mouth, which has always been important, is now the dominant way we learn about the world, with the internet having amplified our chattiness by what, 100x? 1,000x? We're still wrestling as a society with this change (fake news, anyone?), but for marketing, this is the new reality.

As a result, storytelling has come to the fore. At Zuora, we use the Three Rooms mental model of storytelling. You need the story of your product—the how. You need the story of your market—the who. But most important, you need an overarching story that puts your service and your users within a broader social narrative—the why. Most companies (especially ones here in Silicon Valley) have a pretty good grasp of the first two stories. They know what they're selling, and who's buying. They have nice scrolling websites filled with all sorts of product features and client case studies. But lots of them are missing a foundational thesis. A bigger reason to exist. They don't have a why (much less a why *now*). And that's the story you should really be starting with.

In a perfect world, these stories are actually heard in a sequence—the 100,000-foot-high business transformation story, followed by the marketplace story, and only then the product story. You want to create your version of an art gallery, where the goal is to walk the viewer through three rooms in sequence. Room One really isn't about your company at all—it's about the context of your company. It's about what you see going on in the broader commercial world that makes you relevant.

Once you've established the context, only then do you head into Room Two and articulate the value—the objective benefits based on roles and industries. That's when you start drilling a little deeper in order to provide specific role-based advice, industry trends, and relevant case studies. And finally, Room Three is the product itself, like the golden idol at the end of the tunnel in *Raiders of the Lost Ark*. In other words, what your service really does—its features and mechanics.

What you are holding in your hands is the work Zuora does in our Room One, which is the story of the Subscription Economy. We talk about how there's a massive shift going on away from products to services. Even the companies that create amazing physical products, like Nest or GoPro, attach their devices to cloud-based services. We say that consumers have changed in favor of access over ownership, but you know that already, because you've changed yourself. We paint this picture of a rapidly changing world, and then we raise some implications. Advertising can certainly play a role in Room One. It can still be the best way to get broad reach, fast. But those ads have to support a story. So again, what should you do with that promotional budget? Use it to find a story.

So what does your Room One look like?

PRICING (AND PACKAGING)

"Pricing and packaging" is an old-fashioned-sounding term that might remind you of stocking grocery store shelves, but for subscription businesses *it is one of the most powerful growth levers you have*. I can't emphasize this enough. For those unfamiliar with the term, "pricing" means exactly that—the dollar number you assign to the value of your service. "Packaging" refers to the decisions you have to make when associating a specific set of features with a particular pricing plan: what people get for the gold plan versus the silver plan, etc. Pricing and packaging is consistently one of the most popular topics across all of our podcasts, magazine articles, dinner talks, SlideShare decks, you name it. In fact, pricing is the most important of the four P's.

Why is that? Well, pricing for a product SKU is pretty straightforward. Your production expenses and desired profit margins determine your price. In the product world, this is known as "cost plus" pricing. Let's say you build a fidget spinner. You know your manufacturing and sales costs, so you set a profit margin on that fidget spinner. If you sell millions of fidget spinners, you get to discount them in order to undercut your closest fidget spinner competitor and make up the difference in volume. A fidget spinner monopoly is born.

Subscription pricing is trickier. Of course you have costs that you need to account for, but at the end of the day you're not pricing an object, you're pricing an outcome. But how do you express the value in a seat, minute, box, event, you name it? And what do you do about the fact that customers may assign different values to the same outcome? This ambiguity is intrinsic to the subscription model, and it can be either empowering or paralyzing. There's a lot of pressure to not mess it up.

How can you mess it up? Gosh, let me count the ways. You can give your subscription service away for free and then spend years chasing down minuscule conversion rates. You can make things too complicated, with pricing charts that list hundreds of individual features that people are somehow supposed to decipher and evaluate on their own. Or you could err on the side of simplicity with a flat monthly fee, but then you get the Homer Simpson–at-the-buffet problem of having your lunch eaten by people who enjoy your service a little too much. You could tie your pricing to a usage metric that's impossible for your prospects to budget for or anticipate, like selling individual cell phone minutes to a family with a chatty teenager. The list goes on.

But what happens when you get it right? Whoo, boy. Well, customer acquisition becomes much easier, and churn gets reduced. But better yet, as your relationship with each subscriber deepens, as you become a bigger part of their lives, that value is translated into revenue, which you can reinvest in expanding what you do with them, creating a virtuous cycle. You're no longer a victim of circumstance, relying on guesswork and the pricing pages of your next closest competitors. You can create intuitive subscriber journeys that take customers from good to better to best, with relevant incentives and tipping points along the way. And when your pricing model maps to that subscriber journey, this is when (click!) your business model locks into subscriber relationships, and a valuable company is born.

I could write a whole book on subscription pricing (and maybe someday will), but to keep it simple, there are two basic ways to build those growth paths into your service. First, there is consumption-driven growth, which simply means your subscriber is using more of the same base set of capabilities. This is done through pricing. A client business adding more users or storing more data would be examples of consumption-driven growth. I started with the free Dropbox account, but now it's full of pictures of my daughter, and how could I possibly start deleting those in order to stay under the freemium threshold, so I'll start paying for the extra storage, etc. As my customers use my service more, the value of that customer to me increases.

Of course, this requires you to pick the right unit that ties consumption to value. There are a variety of factors to consider here. Some of them might include the need to keep it simple for the user, the ability to translate increased usage into revenue, or the importance of setting a usage "floor" and shaping that usage curve as consumption increases. Next you'll need to consider the unit price, which can be as simple as price per unit, or it could be expressed in usage tiers. Finally, you'll need to look for holes in your model, which usually happen at extremes—someone's using your service too much or too little in relation to your price point. This is where a minimum or base fee can help you set a floor, and a subsequent tiered model can help you reduce the per unit fees at volume. As you can see, you have lots of levers to play with! It's a big formula that you're constantly tweaking—pricing is never finished.

Second, there is capability-driven growth, which lets your subscribers grow into your service by adding more features as their needs expand. This is done through packaging. One common way to do this is to just sell a basic service to the customer, just what they need to get going, and then let the customer add more capabilities over time. For example, if your business is using a customer service application, as you expand overseas you'll want to be able to communicate with customers in multiple languages, which may cost more. We're all familiar with "silver/gold/platinum" tiers of service. These days there are tons of different "skinny" VOD bundles, and "free with advertising" versus paid versions of Spotify, Pandora, and Hulu, for example. But it's important not to drown or confuse people—trapped in that product mindset, lots of companies start introducing a lot of à la carte add-ons, and they muddle that growth path, confuse their subscribers, and often wind up leaving money on the table.

My colleague Madhavan Ramanujam of Simon-Kucher & Partners has an interesting benchmark on this topic: If you have more than 70 percent of your subscribers in your basic package, then you may have a perfectly respectable entry-level service *that will ultimately kill you*. You haven't built a growth path—most of those subscribers are probably happy to stay put. Ideally, if you offer bronze/silver/gold tiers, that 70 percent sits in the silver and gold categories. That means your subscribers are taking advantage of the capability-driven growth path, which also means they're using your service consistently.

Ideally you'll take advantage of both levers, pricing and packaging, because what they essentially represent is increased adoption (consumption-driven) and service innovation (capability-driven). And once you have a steadily growing base of subscribers marching happily along their growth paths? What happens then?

THE GOLDEN AGE OF MARKETING

There has never been a more exciting time to be in marketing. Why do I say that? Because we *finally* have the kinds of customer insights that everyone has been dying to find for the past twenty years. We are swimming in new information. The skills you have as a marketer—storytelling, data analysis, customer knowledge—all of them are crucial to the success of your company. Are you really going to trust the engineers to "growth hack" their way into a great story? They need you!

But here's the thing—once you hit a critical mass of subscribers, and you know who they are, and you see how they're acting, then the job becomes just as much science as art. And that's good news! When the data wonks get together with the writers, that's when the cool stuff happens. Everyone's looking at the same dashboard, and the marketing department becomes a giant test laboratory: spinning campaigns up, surfacing the right narratives, finding the weak spots, and fueling the wins.

Because in the Subscription Economy, there's no more looking elsewhere for answers. No more customer surveys, no more paying for lists, no more waiting six months for a campaign to finish. All the information you need is right there in front of you. Now it's up to you to write the story.

CHAPTER 12

SALES: THE EIGHT NEW GROWTH STRATEGIES

We've all bought something that doesn't work out—that gizmo that sits in the closet for a few years before it gets donated or just thrown into the trash. Maybe it looked cool in the ad. Maybe you used it once or twice, and then the novelty wore off. Maybe there's some planned obsolescence that you can't be bothered to fix. Or maybe there are some things that you buy automatically, that you don't really think about, because you've seen the billboards and the TV ads and the display stands, and when you walk into the store some basic Madison Avenue psychology takes over, and you buy it.

Well, that means "Mission Accomplished" for the companies that sold you that stuff! They don't care about your dispiriting commercial experiences (heck, they don't even care who you are). They got their sale. But with subscriptions, everything changes— particularly with a direct sales force or a team of resellers. You're talking about shifting from an asset transfer model to a long-term relationship. "Relationship" is kind of a strange word to use in a commercial context. Are you really entering into a "relationship" with Netflix? Well, you do, don't you? Sometimes you have great evenings together, and other times you may wonder whether it's all worth it. Look at all the ugly breakups people had with Uber.

No one goes on a "blind date" anymore. Everyone's checking each other out on Tinder and Facebook. But here's the paradox of selling in the Subscription Economy: On the one hand, people already know so much about your company because there's so much information out there. On the other hand, they're more confused than ever before because there are too many choices and too much information. How do you give prospects material that is new and relevant to them, even if they might not even know the right questions to ask in the first place?

We've all sat through the sales pitch that begins with the rep asking you a series of needling questions "so I can better understand your business needs." "Tell me what keeps you up at night." Etc. The obvious intent here is to have you admit to some fundamental flaw or glaring absence that, as it just so happens, the salesperson can solve with their nifty product. The product that has all sorts of bonus features, add-ons, and assorted bells and whistles that your rep proceeds to recite in numbing detail. Maybe that kind of pitch made sense when companies were operating on product margins and competitive checklists, but it doesn't make any kind of sense today. When a potential customer sits down with our company, they're very hungry for information. But at least during the first few meetings, they rarely want to talk about service features or specific use cases. They can walk through our demo or call up some references for that stuff. Instead, they want to hear about two things. First, what are the broader implications for my job and my business if I go with you? Second, and perhaps more important, what are other people out there doing?

So we find ourselves doing a lot of teaching. It's really important for us to be able to say, "We have a lot of customers who are just like you. Before we get into the details, let us share some benchmarks and insights that we've learned from what other companies are doing in your space." Then the dynamic becomes immediately equal. We're having an informed discussion: Our sales team has read your 10-Ks, they've been through your press releases, they've watched your CEO talks on YouTube. Occasionally, we're challenging some of your assumptions and raising those questions that you probably didn't know to ask. The goal is to find alignment, which means that our future innovation, and the company that we will become, is going to be only more valuable to you, not less. As my senior vice president of sales, Richard Terry-Lloyd, likes to say, "With subscriptions, you're never off the hook."

Someone who's about to enter into a relationship obviously wants to learn more about the other side. What's this person's philosophy? Does it fit with mine? Does this look like it's going to be a solid partnership? If I go with this service, am I going to benefit from the collective intelligence of the rest of their client base? Or at least the clients who share my business needs? I know that this service is going to change and evolve—will it fit where I want to be two years from now, five years from now?

At the end of the day, sales is about growth—you're selling a service in order to help your company grow, and your customer is buying a service in order to help themselves grow. And if sales is now tied to establishing and growing subscriber relationships, then the mechanics of growth are different as well. In the old world, you could grow by doing three things: sell more units, increase the price of those units, or decrease the cost required to make those units. In today's world, you have three new imperatives: acquire more customers, increase the value of those customers, and hold on to those customers longer.

There's a common expression: It's way easier to sell to customers than to prospects. I hate that expression because it speaks to a product mindset: you sold them one product, and now you're going to sell them another one. If you're doing it right, expansion—that is, gaining more revenue from dedicated subscribers over time— should happen naturally. When you expand your value, then the commercial benefits will follow. The ability to develop customer relationships over time is where the really amazing subscription companies distinguish themselves from everyone else. If you have a business model that grows as your customers grow, then renewals and upsells will take care of themselves. Hustling add-ons and locking people into onerous terms is a drag and serves only to get people upset.

In working with hundreds of companies, we've learned that the solution to sustaining a high growth rate is to diversify your approach to growth and embrace multiple growth strategies. We've boiled it down to eight essential ones. In any one of our sales meetings, we're talking about at least one of these growth strategies. Let's explore each one of them, and their implications for your sales team.

ACQUIRE YOUR INITIAL SET OF CUSTOMERS

Congratulations. You've come up with a great new subscription offering, and you're ready to unveil it to the world. Perhaps you plan on selling it through a new group of salespeople you've recently hired, or through your company's existing sales force, or through an existing channel of resellers, distributors, and dealers. This is exciting stuff. What's the first thing you need to do? Find the right kind of customers.

Why? Because someday, your future customers will be looking

very closely at your first set of customers in order to gauge whether you're really long-term partnership material. As the saying goes, "You become your customers," so that initial cohort is really important. You really need to take pains to point your sales team toward quality customers at the right price points; otherwise they can wind up like cats who drag back all sorts of strange things to your doorstep.

When we started Zuora, the obvious first strategy was to sell to other software-as-a-service companies like us. But we were always wary of becoming too niche. More important, "flexibility" was a really critical benefit of our service, and we could create that flexibility only by having a diverse set of customers. So we went and found a hardware company, a media company, a consumer subscription company, etc. It made our job harder, but the diversity in those initial customers really set the tone for our business. But they still had to be a good fit! If you chase after anything with a pulse, chances are your initial customers will wind up being terrible. They'll drag your service off in all sorts of unintended directions. Or worse yet, they won't be able to pay much, and your business sinks.

Second, it's really important to avoid the temptation to ramp up a big sales force. Perhaps you are inside a multibillion-dollar company with a large sales force, and you think the best thing to do is to get all of them to sell your service. Bad choice. To begin with, they probably won't know what they're doing. When SaaS was just starting to catch on, lots of traditional on-premise software companies let their people sell both the legacy big-ticket software as well as the lower-priced subscription offering. You can guess which option was more popular with the sales team. In big companies, this is probably the most popular way to kill a nascent subscription service that we've seen. And even if you had a captive sales team selling just your new offering (congrats to you!), you'll want to keep that group small at first. Again, it's a new world—you need to be forever in beta and learn early and often from your first customers. You'll want your sales team to stay very close to that first cohort, rather than bagging a quick commission and moving on to the next hunt.

REDUCE YOUR CHURN RATE

There is a moment in the life of every early-stage subscription business when they have to face down a potentially lethal churn rate. Early on at Salesforce, we had a quarter where we lost more subscribers than we acquired. That was tough. We had the same experience at Zuora. Back in the Qwikster days, Netflix had a quarter where its total subscribers decreased. Everyone goes through it. As opposed to the "WTF Moment," the technical term for this situation is the "Oh Shit Moment."

How can you tell if you're running a successful subscription service? It's pretty simple—you've tamed your churn rate. That marks the transition from adolescence to adulthood, from having a cool new service that people potentially might like to running a mature, successful business. Churn can be countered in part by contract commitments, but lots of subscription companies these days let you opt out whenever you want—that's part of the appeal of their service. No contract length, however, will ever outweigh an utterly maniacal focus on keeping customers happily surprised on a regular basis. There are a number of different ways that companies measure churn, and the reality is that each business is different, with its own unique set of circumstances.

After you get some initial traction, at some point your business will reach a state of subscription equilibrium, a steady state that occurs after your service has been out there in the market for a while, and you've got enough customers to start noticing some trends. At that point, depending on your churn rate, you could be growing, flat, or sinking. If you're losing more customers than you're taking in, then it doesn't matter if you've got a great sales team.

If you find yourself flat or sinking, that's an all-hands-on-deck situation. Don't panic, though. Everyone has to negotiate their own OS moment. But now's the time to ask the hard questions: Are there customers you should not be pursuing (at least not right now)? Are there customers you should fire? What are the real features and usage patterns that really equate to customers finding ongoing value? Are your customers just kicking the tires but never really becoming true ongoing subscribers? Do they need a little push? Could your service be designed or packaged in a different way?

At Salesforce, our core issue turned out to be adoption. It was easy to sell our service, but hard to get people to actually use it (remember, this was in the days when the internet was new—people still used dial-up to get online in hotel rooms, and we didn't all have high-speed access in our homes). We realized we had to teach our customers how to get their people to use the product. Once we solved that, we were back to growing again.

EXPAND YOUR SALES TEAM

Let's say you had a successful launch. You've grown way beyond your first set of customers, you've tamed your churn, and now you're ready to step on the gas. After confirming that your unit economics are healthy (your average customer lifetime value is significantly greater than the cost to acquire and serve) and there's still lots more market to grab, it's time to grow. And growing means expanding your sales team—hiring more reps, getting more productivity out of your current sales team, and signing up more resellers and dealers.

To scale a sales team intelligently, you need to do two things—you need to set up a hybrid sales model, and you need to invest in automation. What do I mean by a hybrid sales model? Well, lots of companies see a self-service sales model and selling through salespeople (commonly called "assisted sales") as two completely different worlds. Self-service is for small businesses, but the really big deals are done through real sales reps, the thinking goes. Don't cross the streams. Well, that's nonsense.

You've probably heard of DocuSign. It's a great company that allows you to electronically sign anything—business documents, bank statements, real estate contracts. DocuSign employs self-service sales to bring on new customers and capture the lower end of the market. For most of its plans—Personal, Standard, and Business Pro—customers can sign up right from the website. Only its Advanced Solutions requires sales support. Similar to Dropbox, the Personal version serves as a lead generation tool. Every one of the up to five documents a user on the Personal plan can send out in a month is branded with DocuSign ("Powered by DocuSign"). This naturally lends itself to viral growth. With more than 85 million users worldwide in 188 countries, this is a guerrilla marketing and sales team unto itself.

Companies like DocuSign that sell to both consumers and businesses often track individuals who sign up with the same domain name (e.g., joe@abc.com and jill@abc.com). Once they see traction from a certain domain, they can then target those customers for an upsell to a business plan. So the self-service channel funnels into an upsell path. Self-service extends to account management as well. Through self-service amendments, DocuSign customers can make changes to their plans. This ability for customers to easily self-service their own accounts is also a great driver for conversions and upgrades. If you do this right, you can keep your sales costs low as you dramatically multiply the "number of feet you have on the street."

The other big challenge as you grow is the amount of paperwork and menial tasks goes up, and the number of mistakes goes up as well. We've all had frustrating sales support calls where the company seems to be resisting our efforts to pay them more for a broader set of services by making us jump through hoops and repeat our address, credit card info, etc. Your sales team needs real-time information on customer subscriptions, billing, payments, and refunds, as well as the ability to automatically calculate proration when customers upgrade to a different edition, suspend a subscription, add more seats, or make other changes. Here is where a "guided selling" model to semiautomate the sales process can be very important. The trick is to find a single architecture that can support self-service, guided selling, and the entire quote-to-invoice-to-cash process. This is something I'll talk more about in the chapter on IT.

INCREASE VALUE THROUGH UPSELLS
AND CROSS-SELLS

You've hit a good stride, maybe you've had another funding round, and it's go go go. So what's next? Eventually every company realizes that the best way to maintain growth is to increase the value you get from your customers. When you can upsell or cross-sell a customer into opting for more of your services, that's a testament to the strength of your relationship. It means you're aligned.

Upselling and cross-selling are frequently conflated, but they are really two distinct growth strategies. While upselling is a strategy designed to sell a more feature-rich (and expensive) service edition, cross-selling is a strategy designed to sell additional services to provide a more comprehensive solution. According to a recent report by McKinsey & Company, subscription companies (with revenue in the $25 to $75 million range) that had the lowest churn were those that cross-sold multiple services to about one third of their customers. The takeaway is pretty clear: the ability to solve for a broad range of customer problems, with a broad range of solutions, increases retention.

Note that this might entail a different sales team structure, with certain people focusing only on new customer acquisition, and other people dedicated to maintaining and expanding existing customer relationships. Naturally, compensation plans enter into this debate: Do you compensate on flat contract numbers, or on the ongoing growth of customer value? If one sales rep inherits a customer that's worth $10 million a year, and another rep inherits a customer that's worth $1 million, how do you think that through? During the early days at Salesforce, we decided to create a team exclusively devoted to managing existing customers. The only problem was that we didn't know what to call them. They weren't customer support. They weren't account management. They were there to help our customers succeed with our service. Marc Benioff suggested we call them "Customer Success Managers." Everyone instantly hated the name. Too bad. Benioff insisted. And today customer success is an entire profession of its own.

Cross-selling also provides an impetus for innovation. Subscription companies that want the ability to cross-sell need to be continually adding new services, features, functionality, and offerings to entice customers to get more value out of the service. Your subscription service is an ongoing experience that can represent wildly disparate margins of financial value. The more you put into that service, the greater the yields. The real work starts *after the sale*. That's why all these companies like Amazon and Netflix keep surprising us with cool new stuff. If you're just managing expectations, instead of creating new opportunities, you're not doing it right.

Given the nature of subscription's recurring revenue business model, it's no surprise that the fastest-growing companies are the ones that are most successful at growing revenues from existing customers. Upselling is critical not only to increasing revenue, but also to customer retention, because the more value your customers get from your services, the more satisfied they are. The trick, of course, is that you need to understand your customers and have deep insight into their usage in order to identify upsell opportunities and build strategic upsell paths.

An effective upsell and cross-sell strategy increases customer lifetime value in the short run while indirectly driving long-term growth. What do I mean by that? Well, in mature subscription services, upsells and cross-sells make up an average of 20 percent of revenues. But additional benefits include lower churn rates, thereby lowering your customer acquisition costs. You need several teams to make it happen. Your marketing team needs to be able to come up with bundles that make sense. Your sales team needs to be able to offer these new options when they're appropriate—giving a subscriber a nudge when they're about to hit a data limit, for example. And your finance team needs to be able to handle the downstream impacts. You also need to be able to measure whether they're successful or not.

New Relic, a leading digital intelligence company, is a great example of a subscription company that has grown very quickly, in large part due to its cross-sell strategy. New Relic is known for its awesome services, with a huge base of devoted developers. It builds on this fan base and drives retention through bottom-up adoption of its services. In other words, rather than focusing its marketing efforts directly at the top (i.e., leadership), New Relic connects directly with developers, providing a wide range of easy-to-deploy services and additional features that solve very real pain points. It does this by offering every service as a month-to-month offering. Any developer can pay a small fee to try out an add-on service without a big commitment. This significantly drives cross-sells. By diversifying its service line and focusing on cross-selling to its devoted base, New Relic not only is increasing earnings per customer and thus overall revenue, but is also poised to grab a larger market share of the global IT management tools market.

LAUNCH INTO A NEW SEGMENT

If your subscription service is designed right, it can go anywhere. It can be universal. The CLEAR expedited airport security service, for example, started off with business travelers, then began selling to families, and from families it started approaching larger organizations about corporate plans. Lots of SaaS companies start off selling to SMB (small to medium-size businesses) before they push into the enterprise. By the way, SMB sales versus enterprise sales could be another way to consider segmenting your sales team.

Box is a great example of a company that successfully moved upmarket. When it first started as a cloud storage and file-sharing company, its revenues through a sales team were less than 1 percent. In other words, it was pretty much a pure freemium service with almost all self-service sign-ups. Now, even though so many of Box's customers are individual users, almost all of its revenue comes from businesses—with the vast majority of its revenue being produced through its sales team.

Box CEO Aaron Levie summed up how this happened: "You

make your service as easy as possible to adopt but you make it so a large enterprise can fully adopt it across their entire company." According to Levie, "There's probably not a single enterprise that we ever sold to that didn't start with users in that organization having adopted Box." When enough individual users within an organization have adopted a service, organizational adoption is the next logical step. Note, however, that unlike self-service routes, enterprise adoption usually begins with a lengthy sales cycle that involves reps, buyers, requests for proposals, and lots of due diligence and negotiating (otherwise known as "haggling"!). But having that process begin with people within an organization who are already using your service is clearly a huge advantage (here Slack also comes to mind).

You can also, of course, launch into an entirely new vertical. As mentioned, we started off selling to other SaaS companies, but now we've got reps who specialize in automobiles, streaming media, IoT, the list goes on. Here's the key takeaway—you need to be able to segment your sales force. You can do it by business size. You can do it by vertical. You can do it by geography. But you've got to do it. Why? Well, at the risk of sounding like Dr. Phil, let's remember that selling today is about building, maintaining, and deepening long-term relationships. You have to know and understand your customers. Monthly box start-ups want to be talked to like monthly box start-ups. Big telcos want to be talked to like big telcos. You have to speak their language, and only a segmented sales force can do so effectively.

GO INTERNATIONAL

Companies typically wait too long to go international. It's a legacy of old thinking. The old way is anchored on geographical and political boundaries. But the world is different now; it's really based on *language*. The reason is pretty simple—the language you use to engage with the internet and your social graph dictates the kinds of results you're going to receive. There are no customs checks when you visit an IP address in Europe. If you're a British newspaper like the *Daily Mail* that specializes in celebrity coverage, it shouldn't be a surprise that 40 percent of your audience comes from the United States. Likewise, if you're selling NBA gear in the United States—guess what? You're also probably doing business in the Commonwealth countries. Are you a streaming video service in France? Well, you've also probably got viewers in the Francophone countries of northwest Africa. It's never been easier to go international. But the main things to consider are establishing some kind of ability to transact overseas, establishing the ability to accept local currencies, setting up the right alternative payment methods, and dealing with price arbitrage. None of these challenges are insurmountable, though, and in general I'd recommend going abroad sooner rather than later.

While going global is a clear growth opportunity for subscription companies, it poses some operational challenges. They're not insurmountable, but to go international, you need to consider three issues: First, the regulatory stuff—business licenses, taxes, data residency requirements. Second, the payments stuff—alternative payment gateways, local currencies, credit cards. Chinese prefer e-wallets, Indians prefer debit cards, South Koreans are more likely to charge smartphone purchases directly to their phone bill, etc. Third, the shop itself—HR, staffing, etc.

Note, however, that you can sell to someone in, say, the United Kingdom, even if you don't have a UK presence. You just send some people over and politely ask your UK customers to transact in your local jurisdiction and pay you in dollars, not pounds. You may get some colorfully worded rejections, but some companies may be cool with it, and at least it's a start. You'll certainly get a sense of the demand in that region. So it's not an all-or-nothing; you can ease into it. The key is to realize that if you are selling in an English-speaking country, you are selling to *all* English-speaking countries.

MAXIMIZE THE GROWTH OPPORTUNITIES OF YOUR ACQUISITIONS

Lots of mature companies eventually hit a stage where they've got significant market share (let's say north of 70 percent), and there are really no more new customers to go get. As a result, growth has to come from increasing your value per customer. This is where acquisition strategies become really important. For those companies that have enough cash to fund an acquisition, it can be a smart move to reinvest that cash into future growth. But access to capital is only one of many requirements for subscription businesses contemplating strategic acquisition.

In addition to cash, an acquiring company also needs a strategic plan that fits in with its business model and day-to-day operations. It also needs the infrastructure in place to support all service lines in one system so that upsells and cross-sells across service lines are workable and the customer experience across service lines is seamless. Successful acquisitions can help a growing subscription business increase its market visibility and market share while enhancing its offerings to build out a more comprehensive solution. You need an integration plan.

You've probably heard of SurveyMonkey. Their management team has skillfully used strategic acquisitions to make them the world's leading online survey platform: Between 2010 and 2015, they acquired six companies—all of which helped them increase market visibility and market share while enhancing their offerings and building out their solution (and folding in the competition). Today, SurveyMonkey cross-sells a number of services that came from these acquisitions. Its first acquisition in 2010 was Precision Polling, which was, according to *TechCrunch*, "like SurveyMonkey for phones." Just a few months after this comparison was published, SurveyMonkey took the hint, acquiring Precision Polling to expand their surveys from online to phone. In 2011 it acquired Wufoo for $35 million, expanding its service line with this easy-to-use solution for building online forms. And it acquired rival MarketTools through a partnership with a private equity firm. This acquisition yielded it three new services, 1.7 million survey users, 2.5 million panel respondents, and some big-name enterprise customers.

SurveyMonkey continued its push into the enterprise market in 2014 with the acquisition of Fluidware, a competitor from Canada with deep-dive features that appeal to businesses. And in 2015 SurveyMonkey expanded into app insights with the acquisition of Renzu and took additional steps to expand its solution with the acquisition of TechValidate, an automated content-generation platform intended to "help every customer now get more from their survey results," according to SurveyMonkey's former CEO Bill Veghte. As SurveyMonkey continues to grow, I look forward to seeing its next smart acquisition. It has clearly mastered the ability to migrate customers from acquired companies onto a single platform for improved accuracy and efficiency of its back-office systems.

OPTIMIZE YOUR PRICING AND PACKAGING

Over the course of the entire lifetime of a subscription business, do you know how much time the average management team devotes to planning their pricing? According to business intelligence platform ProfitWell, the average amount of time a company spends per year on pricing is less than ten hours. That's nuts, especially considering the huge impact that pricing has on your bottom line—it can be much more impactful than similar amounts of effort spent on acquisition or retention.

Subscription businesses need to constantly be optimizing revenue through pricing. In our experience, we see this philosophy reflected by companies that, generally, update their pricing at least annually (which means that they're thinking about pricing constantly throughout the year). Why do you do this? *Because pricing is the key growth lever behind the other seven strategies I just discussed.* I can't stress this enough—your subscription service could be sitting on vast amounts of unrealized value, simply because it hasn't taken the time to test the market with new pricing strategies. You don't need to guess anymore, but you do need to test.

Invoca, a call performance marketing solution, demonstrates this evolved approach to pricing. Because of the nature of its business, Invoca manages multiple pricing dimensions, including minutes, calls, phone numbers, voice prompts, and more. These usage triggers, combined with recurring monthly charges, necessarily create a lot of complexity. The key to Invoca's success is that all of those price triggers *match customer requirements and demonstrate value.* Invoca changes pricing in a central location so that it updates across all systems and sales channels (online, partner resellers, in the quoting tool, etc.). You need to be able to point and click to change your pricing (not rewrite code) and carefully manage the impact of pricing changes for your entire organization so that finance, sales, and operation teams can keep up.

That's it. Those are the core growth strategies. We've talked to thousands of subscription companies in dozens of industries and reduced all their growth tactics down to those eight. As you grow, you'll probably be addressing at least two or three of them at any one time. These days, maintaining a consistent growth rate is an imperative, particularly if you're a software or digital services company. According to McKinsey, if a software company grows less than 20 percent annually, it has a 92 percent chance of failure. Because at the end of the day, it's grow or die.

CHAPTER 13

FINANCE: THE NEW BUSINESS MODEL ARCHITECTS

A few years ago I was invited to participate in an annual off-site for a multibillion-dollar information services company. Now, like almost all "information services" companies today, this one started off as a publishing company. It was founded more than a hundred years ago. Its primary products were professional directories, almost all of which were considered "industry bibles," and so for decades it dominated its space. This company basically printed money. But we all know where this is going.

The internet changed everything, and those professional directories became obsolete as soon as they were published. So this company made a smart pivot to digital and started making relevant acquisitions. And at this off-site, the CEO delivered a really bold vision. He said, "We need to stop thinking of ourselves as a bookseller, or even as a provider of content. We need to start with our customers. What are they really paying us for? What kind of value are we providing them? What do we mean to people? We're going to shift from just focusing on content to focusing on the user experience—become a one-stop shop for our customers' needs, focus on automating their workflows. We've shifted to agile development, and we're rebuilding our entire business model around a customer journey." The effect was spectacular. Everyone in the audience was pumped.

Then the bespectacled CFO took the stage to give the company's annual results. He drily presented the financials for the three core product lines, pointing out that while revenues for all three had declined for the fifth year in a row, the good news was the margins were improving. And he had all the boring charts to prove it. Man, what a killjoy. But sitting in the audience, I couldn't help but observe that something else was missing: the customers! There was no talk about how we are growing customers, which customers are most valuable to us, how our customers are using us, etc. The contrast was jarring. A bold vision for growth was followed by a bland presentation of decline. There was some serious cognitive dissonance in the room.

What a shame. What a lost opportunity. Because if the Subscription Economy is about adopting new business models, then what better function to lead the company through that shift than the finance department?

THE DAY I ALMOST GOT FIRED

But I felt for the guy, because I once had a similar experience. The same thing happened to me. I call it "the day my CFO and I almost got fired." In the early years of Zuora, my CFO—Tyler Sloat—and I presented a plan for the upcoming year to our board. We were excited about our business. Things were going well, and we had an aggressive growth plan that we thought our venture investors would love. We got up there, nailed our presentation, and looked around the room. The reaction was, shall we say, muted. There were several actual frowns. After an awkward silence, one of the board members said, "So, let me get this straight. You want to spend more money, in order to grow less? What is wrong with you guys?"

I'll save you the painful details, but suffice to say the more we dug in, the worse it got. We were probably five minutes from being asked to take our ideas elsewhere. So like a good general sensing defeat, I beat a hasty retreat. We asked for another chance—a mulligan. The board agreed to give us sixty more days to present again. There wasn't much small talk on the way out of that room.

Regrouping, Tyler and I realized that we weren't explaining things the right way. We had delivered the plan using a traditional financial model, a pro forma income statement that was backward-looking and couldn't show the return we would be making on our growth investments. Shame on us for assuming our VCs could do the translation without our spelling it out. We also realized that we needed better benchmarks. But to do benchmarking right, we needed to be able to use publicly available financial statements.

Since I'm still here, you may have guessed that we ultimately recovered. How did we do that? Well, we actually came up with a whole new Subscription Economy income statement. If that sounds like total bunk, just stay with me, but first we have to do some time-traveling. Let's go back around five hundred years. Let's visit Venice.

LUCA PACIOLI AND THE WORLD OF DOUBLE-ENTRY BOOKKEEPING

Our whole financial system today—our ability to generate financial statements, create "books" that can be audited, and compare companies against one another—rests on a concept called double-entry bookkeeping. The basic premise is that your credits have to match your debits. Every time somebody in a grocery store checks their sales figures against the cash they have in their register, that's a form of double-entry bookkeeping.

The guy who first formalized this system, also known as "the father of accounting" (they still have CPA conventions in his hometown!), was a Franciscan friar named Luca Pacioli. He was born outside Florence in 1447 and led a very full life. He was a brilliant, peripatetic man who taught and wrote in Venice, Bologna, and Milan. He wrote a treatise on magic that included the first instructions for card tricks, juggling, and fire eating. He gave lectures on Euclid. He wrote books about algebra, chess strategy, and geometric form and perspective. The latter was illustrated by his collaborator and roommate, Leonardo da Vinci.

Pacioli came from a relatively modest family and was given a vernacular education focusing on trade (the wealthy kids learned Latin and the classics). By the time he moved to Venice to tutor the children of a well-known merchant, he was a sharp, canny young man. This was during the height of the spice trade, when merchants from the Venetian city-state journeyed to the Middle East and Asia to source rare incenses, herbs, and opiates. Because of the time and distances involved, they often dealt with credits and debits, and it was easy to make mistakes, to lose track of who owed what to whom. Early versions of double-entry bookkeeping date back to the 1300s, but Pacioli was the first to formally codify it. His book was modestly titled *Summary of Arithmetic, Geometry, Proportions and Proportionality.*

Here was Pacioli's great maxim—you don't go to bed until the debits equal the credits. For every financial transaction, both column entries have to match. When you record these entries accurately and punctually, you have an audit trail, which lets people get paid fairly. You also get an aggregate view of your total assets, which are equal to the sum of your liabilities and your equity. It's a simple but powerful equation that lets you build things like income statements, balance sheets, and cash flows. In fact, here's a typical income statement that they teach in any Business 101 class (in millions):

Net Sales	$100
Cost of Goods Sold	(40)
Gross Income	$60
Sales & Marketing	(20)
Research & Development	(20)
General & Administrative	(10)
Net Income	$10

Pretty straightforward, right? This says you sold one or more units of a product at a total sale of $100 million. The income statement brings out the fundamental costs that went into that unit. There are four of them: How much it took to make that unit (cost of goods sold), including the raw materials, production costs, etc. How much it took to sell that unit (e.g., commissions to salespeople, money for the channel). How much R&D the company is spending to create the current unit. And finally, how much overhead (e.g., finance, HR, executives) management needed to run the company. As you can see, some of these costs are fixed, like how much R&D you need, so the more units you sell, the lower your per unit cost. That's why they call them "accountants"—it's their job to account for all the costs that go into making the unit of sale. Congrats, I just saved you a hundred grand in MBA tuition.

But let's go back to the day I almost got fired.

Shortly after Tyler and I presented our upcoming plan to the board, we realized that this model may have worked fine for the last five hundred years, but it was entirely wrong for a subscription business, for three key reasons. First, the traditional income statement does not differentiate between recurring and nonrecurring dollars. It's like saying there's no difference between a dollar and a dollar that keeps happening every year for the next ten years. Recurring revenue is the cornerstone of subscription businesses, but traditional accounting concepts were never designed to recognize this fact. Second, sales and marketing is matched to past goods sold. It's essentially a sunk cost. I'll get into this later in the chapter, but subscription businesses need to think strategically about sales and marketing spend going toward driving future business. And finally, *this is a backward-looking picture*—it's all about money already earned, expenses already paid, actions already taken. Subscription businesses are all about forward visibility: how much money I know I can count on over the next twelve months, so I can account, plan, and spend accordingly.

So we decided to come up with something new.

INTRODUCING THE SUBSCRIPTION ECONOMY INCOME STATEMENT

Every smart subscription business I know focuses on something called ARR, or annual recurring revenue. What is ARR? Simply put, it is the amount of revenue that you expect your subscribers to pay you every year. It is the revenue that recurs, as opposed to the revenue that gets booked only once. Every quarter, subscription businesses look at how much their ARR has grown, using the following formula:

$$ARR_n - Churn + ACV = ARR_{n+1}$$

You start the period at a specific annual recurring revenue run rate.	You spend a percentage of that ARR to service the base (COGS, G&A) and to invest in R&D.	In doing so, you make every effort to minimize the loss of ARR.	You also invest to grow your ARR by acquiring new Annual Contract Value, from both new and existing customers.	You then arrive at a new ARR run rate as you begin the next period.

Based on this formula, Tyler and I constructed a Subscription Economy income statement, one that better reflects how we thought about the business. This is what we presented to the board sixty days later. Since then we've shared this model with countless other companies, as well as a number of financial analysts who cover subscription businesses.

Given the formula above, a Subscription Economy income statement should look more like this (revenue in millions):

ARR	$100
Churn	(10)
Net ARR	90
Recurring Costs:	
Cost of Goods Sold	(20)
General & Administrative	(10)
Research & Development	(20)
Recurring Profit	40
Sales & Marketing	(30)
Net Operating Income	10
New ARR (or ACV)	30
Ending ARR	$120

Let's break this down.

ARR—Notice again that instead of starting and ending with income, now we're starting and ending with ARR, or annual recurring revenue. Again, that's because this income statement looks forward, not backward. Whereas a traditional statement says "you made this much revenue in the past quarter," this statement says "you are starting the quarter with this much recurring revenue." That's a huge quantitative difference. ARR is known recurring revenue that you can bank on.

Churn—Sadly, not all recurring revenue actually manages to recur. Even if you've got the best offering in the market, you'll still have customers who leave you: If they're a consumer, they might get bored with your offering or switch to a competitor who's offering them a shameless discount. If they're a business, you might lose your advocate, or they might get acquired or go under. Other reasons, of course, are more self-inflicted: poor adoption or utilization, product gaps, lame consumer marketing, a lack of resources and expertise. Regardless, this is called churn and is a reduction of ARR. So instead of being able to bank on $100 million, in this example the company subtracts a projected churn of $10 million to get a net ARR of $90 million. This is what they can bank on. Now, how will they spend that money?

Recurring Costs—The first question is: What do we need to spend on to service that ARR? After all, your existing customers are expecting a service! Here is where we've rearranged some costs. We've pulled up COGS, R&D, and G&A to be "above the line," in the sense that these are the costs you need to spend to service your ARR. Now, some companies or analysts get fancy and try to allocate only some percentage of these costs as what is needed to service the ARR, but we decided to simplify things and assume that COGS, R&D, and G&A are all associated with servicing the recurring revenue. Hence, they are recurring costs. The benefit of this assumption is it makes it easy to benchmark against other public companies by using their available financial data.

Recurring Profit Margin—Your recurring profit margin is simply the difference between your recurring revenues and your recurring costs. This number gives you the intrinsic profitability of your subscription business, as there is certainty in your recurring revenue, as well as certainty in the costs to service that revenue. In the example income statement it's $40 million, which represents a very healthy margin. There's a lot of hand-wringing around how profitable subscription businesses can be. When we look at a subscription company's financials, Tyler and I always look at the recurring profit margin first to get a sense of how strong the business truly is.

Growth Costs—So what about Sales and Marketing? Here lies one of the biggest differences between traditional and subscription businesses. In a traditional business, the cost of sales reflects how much I spent to get that dollar of revenue. But in a subscription business, sales and marketing expenses are matched to future revenue. Why? Because the sales and marketing I spent this quarter adds to the ARR, but the revenue I will see from that ARR growth will come in future quarters. In traditional accounting lingo, your sales and marketing now acts more like a "capital expenditure," or capex. Essentially, these are costs you spend to grow the business, either from existing customers or from acquiring new customers. That's why we call them growth costs. Again, for simplicity and ease of benchmarking, we assume that 100 percent of a company's sales and marketing costs are spent in order to grow.

GROWTH VERSUS PROFITABILITY

So now you can see the relationship between the recurring profit margin and growth. The higher your recurring profit margin, the more you have to spend on growth. In the example on page 180, the company actually chose to spend almost all its recurring profit margin on growth and wound up 20 percent bigger at the end of the period.

At this point you may be asking, why not spend *all* of the recurring profit on growth? Why not, indeed? If you believe you have a big potential market and have control over your churn, you can run this play year over year, and you're growing by 30 percent annually. And when the time comes to finally start taking profits, you're working off a much bigger recurring revenue stream.

This is why a lot of subscription companies may "look" unprofitable, but they are actually awesome businesses. It's also why I roll my eyes every time I hear an analyst complaining about the lack of profitability at companies like Salesforce and Box. Even while the Subscription Economy has taken hold across multiple multibillion-dollar industries, investors, analysts, and investor media continue to miss the fundamental differences between product and subscription companies that make their financial measurements just as disparate.

When I was at Salesforce, we spent a lot of time and energy educating investors and analysts on the vast performance differences between subscription software companies and traditional software companies. Lots of them remained fixed on the price-earnings ratio, and could not fathom investing in a company trading—at that point—200x future earnings. We knew that operating profit was essentially meaningless to measuring our value.

Honestly, as an investor, I would ding a subscription business that brought operating profit to the bottom line, seeing it as a signal from the company that it's cutting sales and marketing spending because it can't efficiently acquire new bookings! Here's the key takeaway—*it is perfectly rational for subscription businesses to spend all their profits on growth, as long as their bucket doesn't leak.*

Remember, as long as you are growing your ARR faster than your recurring expenses, you can step on the gas. As Ben Thompson of *Stratechery* notes, "You're not so much selling a product as you are creating annuities with a lifetime value that far exceeds whatever you paid to acquire them."

TYLER'S SLIDE

Below is a slide that my CFO, Tyler Sloat, presents to the company every three months, when we do our quarterly all-hands. He also presents it at every public speaking opportunity he gets. He shows this slide to bankers, analysts, the press. He could talk about this slide in his sleep—sadly, I'm sure he does sometimes. Everyone at Zuora has it stamped into their brain. We call it Tyler's Slide:

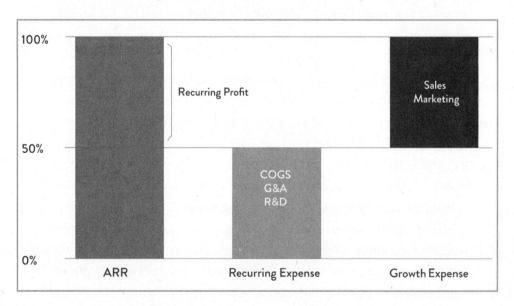

Now why is this chart so important? Because this is how Tyler guides our whole company on how to be a great business. We've seen how the financial model behind all these subscription businesses can be so powerful. It's why Box, despite all the hand-wringing by analysts during its IPO about its profitability, has seen its revenues grow by more than 100 percent since the IPO. This is the opportunity that publishing CFO was missing out on—the chance to guide his business to become a stronger, more competitive, more sustainable business.

So how does Tyler use this chart to do this? Well, the first thing you notice is that all the elements of the Subscription Economy income statement are there: ARR, recurring expenses, growth expenses, recurring profit margin. And notice how everything is shown in relation to ARR. So in one simple image, we see it all. Now, there are three ways we use the chart to drive our business.

First, we use it to manage our recurring expenses. In other words, we use it to budget. Every year, we determine COGS, G&A, and R&D spends as a percentage of our ARR. Budgeting used to be a total nightmare, with all sorts of politics and loud voices. Now it's actually pretty simple. All our department heads know that the best way for them to grow their budget is for them to contribute to growing our ARR. How are they using their resources effectively toward that single goal? When ARR grows, your budget grows.

Second, we use it to manage the trade-off between recurring expenses and growth expenses. You can do three things with those two bars on the right of Tyler's slide. You could keep them at 50 percent each and still grow at a respectable rate (remember that hypothetical income statement has us growing at 20 percent). You could increase your recurring expenses and decrease your growth expenses (let's say 70 percent recurring/30 percent growth), which would help your bottom line. Or vice versa: you could increase your growth expenses to 70 percent and decrease your recurring expenses to 30 percent, and spend more in order to grow faster.

We've frequently decided to keep our recurring expenses flat, even as our ARR is growing, and tilt our spend toward growth instead. Why? Well, if you can demonstrate that you know exactly what your return on growth spend will be, you can tell your CTO that instead of hiring five new engineers this year, she'll be able to hire twenty new engineers next year. We've frequently made that offer to our department leaders—if you can stay lean for three quarters, for example, the dividends we generate from our business model will be worth it. This is also why recurring profit margins are so important for subscription businesses in terms of their ability to grow.

Third, we use this chart to govern our sales and marketing spend with a concept called the Growth Efficiency Index, or GEI. Let's say you spend $1 on sales and marketing—efforts intended to grow new business. How many new recurring revenue dollars does that buy you? This is your Growth Efficiency Index, and it tells you how much and how fast you will grow. If your GEI is greater than 1, you're currently spending more money on new business than you're taking in. If your GEI is less than 1, you're spending less than you're taking in—clearly where you want to wind up! Lots of companies that are chasing growth, however, frequently have GEIs in the 1.0 to 2.0 range.

As a matter of fact, you can increase your growth spend to the point that you are actually losing money. Come again? Well, if you have cash in the bank and access to capital, if your recurring profit margin is healthy, and your churn and GEI numbers are healthy, you should step on the gas. It's a no-brainer. It's just good for the business. You're spending money on acquiring recurring revenue, so you don't have to spend money to renew it every year. Amazingly, some folks on Wall Street still don't seem to understand this.

BEYOND COUNTING BEANS: IT'S TIME FOR FINANCE TO LEAD

A few years ago, Tyler started hosting an annual get-together of CFOs in Half Moon Bay. Initially, I was pretty suspicious. What were they doing? Geeking out on the newest FASB bulletins? Playing fantasy baseball? As it turns out, they were swapping notes: here's what we're tracking, here's what we think is important, here's what we think is overhyped, etc. Every year, that group grew bigger and bigger. Today, Tyler hosts more than a hundred CFOs, finance execs, bankers, and analysts at *Subscribed* San Francisco. They talk about the market, go over survey results, and dive into best practices.

If there's one lesson I've learned from attending every year, it's that the job of the finance team has changed dramatically. Way back in the twentieth century, the mandate of your finance department was to track all the expenses in the company and map them to the unit sale: What was the marginal cost to make that product? What was the cost to come up with the original idea? How much money did we have to give to our channels as part of that sale? How much overhead do we have to manage? Traditionally, 80 percent of a CFO's job was to tell people what happened. To keep score. To track the budget. The other 20 percent was to interpret those numbers in order to direct resources, create forecasts, and manage strategy, to write the next part of the story. Today that ratio has flipped.

Now, compliance and reporting are still important, don't get me wrong. Those are table stakes. But particularly since the economic downturn in 2008, CFOs have found their responsibilities dramatically expanding to respond to new market challenges and a rapidly evolving regulatory environment. It's a new world built around dynamic business models, and increasingly, finance teams are the ones taking charge. Because at the end of the day, a business model is different from a budget. A budget is about handing out head-count and expense against assumed revenue. A business model is a mix of strategy, insight, and ideas that winds up as a quantitative but fluid framework—it both effects change and reacts to it.

And increasingly, that model belongs to the finance team, the new business model architects.

CHAPTER 14

IT: SUBSCRIBERS, NOT SKUS

Everybody loves to complain about their IT department. IT is a drag, a bottleneck, etc. Of course that's all nonsense. When you really think about it, the last twenty years of IT has been pretty amazing. If the overall goal of an IT department is to increase a company's efficiency by standardizing its business systems, then most IT departments have been remarkably successful at that effort. Back in the nineties, they installed big ERP systems that finally gave their companies legitimate systems of record: Oracle, SAP, JDE, PeopleSoft, etc. Everyone was happy (well, mostly everyone). All of a sudden, IT got cool.

Then this thing called "the cloud" happened—suddenly, there was an explosion of new plug-and-play SaaS business applications everywhere. At first, IT departments eyed this stuff warily, since the idea of businesses buying all these applications right off the shelf felt sketchy. But once their security issues got resolved, they realized that all these SaaS applications made them much more responsive. Need an easy-to-use expense management app? I'll whip up one of those right away with Concur. Need a marketing automation tool? Marketo has you covered. Want to share and se-cure files more easily? Box to the rescue. Can you believe it? IT got even cooler!

But lately, I grieve to report, there have been some bumps in the road. IT is starting to drag again. Businesses are starting to ask questions that IT can't answer. Why? Because the foundational IT systems of these businesses are based on stock keeping units (the identification codes assigned to product lines), not subscribers. For example:

Who are my subscribers? Try asking SAP or Oracle how many active subscribers you have at any one time, and they'll be stumped. The concept simply doesn't exist in their universe. Systems of re-cord for orders, accounts, and products? Sure. But ask your ERP how much upsell business you've done, or how many customers have renewed in the past year, and you'll get a blank stare. ERP is simply not built around customer-centric transactions. And unless you can monetize customer relationships over time, you're dead in the water.

Can I price this service the way I want to? A few years ago I was speaking with the digital lead at a large, respected newspaper, one of the first to establish a successful paywall. She told me they had offered only two pricing plans over the last ten years: standard and premium, both annual plans. They knew they could do better: what about a day pass for casual browsers, who only wanted to read a few articles? Sounded like a good idea. So they spent six months and lots of effort with the IT team to launch this capability, to huge fanfare. And what happened? Turns out it didn't work. Which is fine in theory—it's always good to experiment, but not if it takes that long. That's when they realized that the only way to figure out the right pricing was to constantly try new ideas, and that six months to try out an idea simply wasn't acceptable.

It's really hard to experiment and try new things when all those experiments cost so much in terms of time and effort. Subscription services run the gamut from simple monthly recurring charges, to usage-based charges, to one-time charges, to "all of the above." Meanwhile, a single price change in an ERP system can take months to implement. In the old world, if you wanted to raise a monthly price from ten bucks to eleven bucks, for example, you'd have to wait a long time to find out if that was a good idea or not. That's no longer viable. It's imperative that subscription companies have the ability to conduct rapid A/B price testing when trying to gauge appetite for a new offering.

Where's the "Renew" button? At their core, ERP systems give you only a "Buy" button for tracking transactions. The problem is that subscriptions are constantly changing, as customers sign up, upgrade, add on, and renew their service. Unfortunately, ERP systems force companies to resort to hokey workarounds to get their pricing right, like creating a different SKU (or entry in the product catalog) for every month of the year, just so you can process monthly renewals. They're missing the critical tools you need to process a subscription life cycle over time. Let's say you're a language learning app that wants to introduce a new subscription service for executives who travel frequently. What would your ERP call this new service? Probably something like "February Service SKU." Painful.

Why can't I sell to everyone? Let's say you're a manufacturing company that successfully launches an IoT analytic service targeting big enterprises. So far, so great. But now that you're selling digital services, you should be able to sell to small businesses and consumers as well, right? Not if your ERP system charges you $400 every time you need to send out a bill! Subscription Economy companies like Salesforce and Box have found success by selling their services to everyone from individual users up through very large enterprises. They need tools for managing things like high-volume recurring payments in the B2C world, as well as tools for managing high-complexity invoices and contracts in the B2B world. And those tools need to manage customers who may come through different channels such as Web self-service, mobile devices, direct or channel sales, or Facebook. Legacy enterprise technology makes you choose between B2B and B2C, when what you really need is the ability to sell B2Any.

What's going on with my financials? Here's another true story—a friend of mine took over as CEO of a SaaS company. He decided to run a little experiment and asked both his finance team and his sales team for the company's latest monthly recurring revenue numbers. He got two wildly different figures. And as it turns out, the sales team numbers were more accurate! Huh? Subscription businesses live or die by their ability to measure the ways that bookings, billings, cash flow, and revenue are interrelated. Unfortunately, this data lives in different software silos. Bookings fall into CRM, billings and cash flow live in your General Ledger or ERP system, and revenue is too often calculated in a series of complex spreadsheets. Good luck stringing all of that together.

The list of unmet needs goes on and on. All of a sudden, IT is falling behind again.

Why is this happening? Well, the nature of ERP systems is part of it. They were made for tracking products on pallets, not sustaining services that are consumed over time. Remember "any color as long as it's black"? That's ERP in a nutshell. After decades of focusing on standardization and helping their company scale up product businesses, IT is waking up to realize that those same systems are now too restrictive for the dynamic nature of the new subscriber-centric business models. Let's look at what a typical IT architecture looks like today:

LEGACY IT ARCHITECTURE

CRM + Linear Quote-to-Cash + ERP

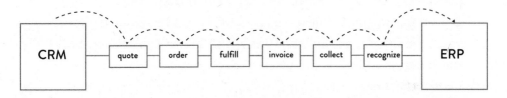

There's your financial system on the right. This is what most companies put in fifteen to twenty-five years ago (these systems are almost all owned by Oracle now). Shortly after they got ERP systems, most businesses settled on a CRM system like Salesforce or Microsoft. So your CRM and your ERP became the two anchors of IT infrastructure. But then, of course, you have all kinds of stuff in the middle. You have quoting systems, ordering systems, fulfillment systems. And they're all daisy-chained together in a strictly linear format. They're serialized. And what ties them together is the almighty "order."

To understand how all this works, let's visualize the old departments that these systems were designed around. Let's say you have a customer who wants to buy 50 units of your widget. Your sales department would issue a quote that says, "If you want to buy 50 widgets, that'll cost you $10,000." The customer agrees to the price, and so that quote gets carbon copied (for all you nonmillennials who remember carbon copies), and a slip gets passed over to the next department, the orders desk. Then the person at the orders desk says, "All the rules are being followed here, and the customer passes the credit check, so I'll go ahead and submit the order." And then another slip of paper gets passed to the fulfillment desk (you can see where this is going).

So then the friendly fulfillment guy down in the warehouse puts 50 widgets in a box and tapes it up and ships it. Then another piece of paper gets passed to the inbox of someone who calculates the product costs and taxes and shipping costs and any applicable discounts and mails out an invoice, which then flags the collections department, which sends out the PO and charges the credit card. Finally, an accountant in the fulfillment department books the ten grand. Maybe the piece of paper moves through a pneumatic tube, I'm not sure.

Now the reality, of course, is things are much more complicated. Any organization of any decent size will not have one of each of these systems; they'll have dozens. Whether they're the result of acquisitions or new business units, these fulfillment chains tend to multiply. Maybe you put in a new quote-to-cash system for a new product line. Maybe you inherited one through an acquisition, so you kept it. Maybe you added a whole new distribution channel—value-added resellers, digital ecommerce, whatever—that needed its own system. Maybe you launched in a new country, which required a different system that spoke to the local regulations and payment protocols.

I recently talked to one $5 billion company that was simultaneously running forty-four discrete quote-to-cash systems. Just writing that makes my head hurt. I talked to another company that had a system built in the late seventies that could be administered only by the person who originally designed it (they gave that person a free gym membership for life and would not let him retire). You can see how it happens, though. You set yourself up with a big ERP installation because you want to efficiently ship and sell a product. Over time you put in a quoting system for your sales team. You proceed to buy several more companies, each with its own subsystems, and figure that it's way easier just to change the stationery and the email signatures than all the back-end stuff. Then one morning you wake up and no one knows where anything lives, and it takes six months to get anything done.

In the old world, the system kind of worked. The carbon copies passed through all the slots, and things got done. But we're not in that kind of static, transactional world right now—things are much more dynamic. Subscribers are constantly upgrading, downgrading, suspending, or adjusting their services. They go on vacation overseas (suspension) and need a new cell service (add-on). They're constantly bombarding these poor linear systems with new requests and changes. Companies need to have the agility to respond to thousands of events a day, but they're also restless. They want to experiment with go-to-market strategies, pricing models, and new services because there are as many threats as opportunities out there.

So what happens when this linear order-to-cash system meets this new dynamic world of services? What happens when the unstoppable force meets the immovable object? It's not pretty:

SUBSCRIPTIONS + LEGACY IT ARCHITECTURE = CHAOS

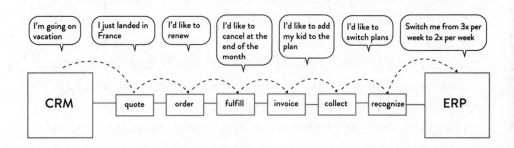

These systems have three big problems—three "can'ts." First, when you want to change your subscriber experience, you can't. You're forced to recode each system to handle a never-ending wave of customer events. For example, what happens when you want to change the pricing and packaging of your service? It's not as easy as going from a dollar to a dollar fifty. Say you're switching from weekly to monthly service contracts, or switching pricing models from seats to usage. Your nice little linear model doesn't respond well to change. Change one thing, change everything. Since everything is daisy-chained, if you make a quoting change at the top of your system, everything else gets affected in turn. After a few years of this, you're left with all sorts of strange hacks, skunkworks, and scary closets.

Second, when you want to change your pricing quickly, you can't. A CFO of a $300 million software company once told me: "We've got great customers—they tell us what they'd like to see next. But our biggest challenge is always pricing and packaging. Even though we think we can make millions of dollars with a new capability, we'll just throw it into our existing product for free because we don't know how to price for it. It takes too long to figure out."

Third, let's talk about business insights. When you need to get a single view of your customers and their entire subscriber life cycle, you can't. Everyone wants to be a "big data" company these days. They want a 360-degree look at their business, so they take all their subscriber and financial information and dump it into a "data lake." Sure, throw it all into the lake; everything will work itself out. We'll get the insights we need. The lady in the data lake will tell us! But in the process, you're basically taking information from potentially dozens of systems and throwing it into an undifferentiated mess. You're asking a jigsaw puzzle to solve itself. These are disparate systems that were never designed to work together.

And what happens when these linear order-to-cash systems get asked to report subscriber data and recurring business metrics? Tragedy strikes. Your IT group has spent all this time scaling your back-end operations to be efficient, but in the process, of course, they've actually made things *more* vulnerable, resulting in terrible customer experience, unnecessary overhead, increased compliance risk, stalled innovation, and stunted growth. You're stuck.

So what's the solution? IT organizations are realizing that they need to evolve their architectures to meet the new needs of their organizations. And what should the new architecture look like? Well, it should look like something that puts subscriber IDs at the center of your architecture. Remember my favorite diagram at the beginning of the book? The one you won't ever forget? The new IT architecture should look a lot like that. It should look circular, not linear:

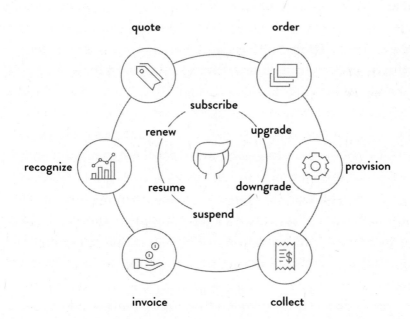

Subscriptions are an ongoing dynamic cycle of actions: renew, suspend, upgrade, downgrade. So the subscriber actions on the inner circle of this diagram need to inform the business systems on the outside. Let's go back to subscriber experience. Let's say a subscriber hits a threshold, so that triggers a prompt, a credit check, and potentially a new usage tier. Or let's say someone lands in another country, so that activates a roaming service and an entitlement check. You need one place where you can coordinate all these if/then scenarios.

What about pricing and packaging? You need to be able to spin up new services and pricing models in minutes, not weeks (the same way *Financial Times* took advantage of Brexit weekend). Almost 80 percent of the companies we work with use some kind of usage-based model, so you need a system that lets you experiment with tying different prices to different value metrics: seats, minutes, boxes, events, gigabytes, locations, texts, family members, you name it. When you're launching a new service, now many of those "outer" systems serve as touch points that inform "inner" subscriber actions. You're studying those reactions (let me upgrade, let me downgrade, get me out of here) and iterating accordingly.

Finally, how about business insights? How does Netflix know they can spend $8 billion on original content next year? Because they can report on their business with the metrics that matter— we've got these recurring costs, we've got these recurring profits, so we've got this margin to do with as we please. How is Slack able to give companies credit for users that later become inactive? Because they have a direct line of sight into their subscriber base, and they know that by aligning their service more directly with their customers, and not charging for empty seats, they win in the long run.

The old daisy-chained systems just don't make sense anymore. As Aaron Levie of Box once noted, you used to be able to do fine just by deploying SAP better than the next guy. That's no longer the case. Today, IT is where you *compete*. It's where you spin up new services, new experiences. It's where you set up test beds and experiments. It's where you iterate and scale. It's where you find the freedom to grow. And more and more business systems are enabling this kind of freedom because they're based on subscribers, not SKUs.

CHAPTER 15

BUILDING A SUBSCRIPTION CULTURE WITH THE PADRE OPERATING MODEL

Congratulations—you've finally launched your new subscription offering. Your engineers are experimenting in beta, your marketing team is reinventing the four P's around your subscribers, your sales team is finding a clear path to growth, finance has come out of the back office and is driving the company's business model transformation, the IT team is launching new services and iterating. The behavioral data is starting to stream in, as well as the recurring revenue. Things are awesome.

Or at least, they're supposed to be, but they're not. Things are very far from awesome. Sales is pestering the engineers about which new features to build. The engineers are tersely suggesting that the sales team stick to sales. Finance is telling sales that a lot of these new customers are churn risks. Sales, as always, takes it out on marketing. And everyone dumps on IT, all the time, for every reason. Outside of raising red flags and putting out fires, there doesn't seem to be much interdepartmental cooperation going on.

So what's happening? The business model has changed, but the old way of doing things hasn't.

A SUBSCRIPTION CULTURE

It turns out that being a "customer-focused company" is a simple concept that is, in fact, very difficult to realize. It requires a cultural change. Product cultures are built around thinking and organizing like assembly lines: stay in your lane, do your job, then pass it on to the next person. That no longer works. Subscription cultures are about making sure your customer continues to succeed with your service over time, and translating that ongoing value into revenue. And why is creating that subscription culture so hard? Because the same organizational structures that worked in the past are now hindering the future. Organization is the biggest thing holding us back—just like in the horror movie, the scary phone call is coming from inside the house.

We're all taught the same basic functional structure: product, marketing, sales, IT. That structure made a lot of sense when those large postwar corporations I mentioned in the introduction were trying to grow quickly. The thought was that strict departmental organizations scale much faster. But today's customers are telling you to do things differently, and they couldn't care less about your org structure.

These were the issues we were facing during the early years of Zuora. The business was scaling nicely. But like so many businesses, we were starting to get siloed. Increasingly, our departments were arguing, and we would argue about who should do what—instead of what was good for the customer, who was increasingly getting lost in the noise. It was clear we had to break down the siloed thinking that was starting to set in. But we weren't really sure how to do it.

Our first idea was to organize by customer teams. If we were going to be a truly customer-focused company, then every customer should have their own dedicated team with complementary skill sets, right? So you're the marketing person on this customer. You're the salesperson on this customer. You're the product person on this customer, and so on. But that turned out to be a pretty problematic approach. It wasn't flexible enough. And it didn't really scale well.

Besides, there are still benefits to groups of shared expertise sitting together. You want developers to sit with developers because they make each other better. You want salespeople to be sitting with salespeople, swapping notes and sharing common training. Sure, it's not like there's no more skills specialization, but a functional organization that only looks at departmental responsibilities. We wanted clear functional roles, but also a way of looking at our company that transcended org charts and put the customer squarely in the middle. We also wanted every employee to feel a sense of ownership over the entire company.

That's why we invented PADRE.

INTRODUCING PADRE

What's PADRE? Well, it's our way of visualizing our company as an integrated organization composed of eight subsystems, all tied to the customer.

We start with Pipeline, which a consumer-focused subscription company might call positioning. The key goal of the pipeline subsystem is to build market awareness and translate that into demand. You do that by getting the marketplace to understand your story: who you are, why you exist, what you do, what benefit you provide. You're also not just addressing potential customers, but

202

the people who influence them: journalists, analysts, the adjacent vendors they interact with, etc. The idea is to get an interested (and, ideally, well-informed) potential base of subscribers to engage with the company: visit the website, download the app, talk to your salespeople or resellers.

Then we have the Acquire subsystem, which encompasses the so-called buyer's journey. How does this potential subscriber make decisions? What are their criteria for success? What are their alternative solutions? What are their potential objections? Who do they have to check in with: their spouse, their family, their boss, their CFO, their team? How do we make sure we're aligned so that the better they do, the better we do? Only once a company becomes a customer and enters into a contracted subscription relationship with us can we actually start creating value. And the more customers we have, the more we can learn about how subscription models work in various industries, and the more best practices and benchmarks we can share with everyone. So how do we make that happen?

Deploy. How do we get our customers up and running as quickly and efficiently as possible? Because if we don't get the integration right, we're pretty much doomed from the start. Did you wear your Fitbit for a couple of days and then forget about it? Did you set up your village in *Clash of Clans*? Is your sales team actually using this new content enablement system? Did you enjoy the experience of cooking that first Blue Apron or Sun Basket kit? The key is to get your subscribers up and running, so they can quickly get invested in your service.

Run. As a subscription company, you succeed or fail based on how well and how long your subscribers take advantage of your service. Anything you do that doesn't feed into the customer success flywheel is detrimental to the growth and value of your business. Are your subscribers logging in every day, every week? How do you make sure they're successful on a daily basis? What kinds of features are they *not* using? Are they happy with performance? Are there any "downtime issues"? Are they giving you decent grades? If something bad happened or somebody messed up, how did you learn from it?

Expand. You want three things out of your subscribers: retention, growth, and advocacy. They have to get value out of your service and make sure you're delivering something that's better than what they can get anywhere else. How do you do that? How do you deepen the relationship? Are there other features you want them to use? If you like Uber, have you tried UberPool? If you're listening to Spotify on your laptop at work, have you streamed it on your Sonos? And so on. How do we make sure that customer success isn't just a functional role, but a guiding principle of our company?

PADRE is also supported by three core subsystems that manage the "back of the house": People, Product, and Money. We need to hire great people and help them develop and grow. How do we make sure our headcount plans square with our business goals? We also need to build a great product that helps our customers move beyond solving inefficiencies toward creating new opportunities. How do we continue to research and develop great ideas in a sustainable way? Finally, we have to manage our resources effectively. How do we spend our money in a way that makes sense for how fast and how big we want to grow?

So the full name is actually PADRE/PPM, but we just say PADRE for short. Here's what it looks like:

POSITION

- Web & Social Media
- Public Relations
- Events

ACQUIRE

- Sales Teams
- Reseller Channels
- Self-Service

DEPLOY

- Implementation
- Customer Training
- Customer Adoption

RUN

- Account
  Management
- Technical Support
- Customer Success

EXPAND

- Increased Consumption
  (Upsells)
- Increased Capabilities
  (Cross-sells)
- Customer Evangelism

PEOPLE

- Recruiting
- Onboarding &
  Training
- Career Development

PRODUCT

- Research &
  Development
- Product Marketing
- Beta Innovation

MONEY

- Finance
- Operations
- Legal

Now hold on a minute, Tien, you might say. Isn't this just a re-branding exercise? Isn't this the same rigid organizational structure, just with a new name? Isn't pipeline just marketing, and acquisition just sales, etc.? To which I would respond: Hell No. This is a perspective that transcends departments, roles, and org charts. It's an enduring way of looking at a subscriber-focused company. Things change all the time around here. That's natural. Priorities shift. People arrive, people leave, working groups form, projects finish. As we grow larger, we start specializing, and things by definition get more complicated. The organization may start to grow and branch out in all sorts of unexpected and unusual ways, *but these eight subsystems stay the same.*

And here's the most important part—the only way these individual subsystems can succeed *is with cross-functional coordination.* Let me give you an example. We were having problems with deployment—getting people up and running on our system. We manage a pretty sophisticated financial enterprise software platform, so it's not as easy as sending someone a password and a log-in. So the first instinct would have been to ask the deployment group (in our industry this is called "professional services") about the problem. But we didn't do that. We asked everyone about the problem.

We had an off-site meeting where every department had to brainstorm ways to improve the deployment process. And we discovered that the problem had a lot to do with the fact that our sales group wasn't setting proper expectations. So they said, "How about this—from now on, we're going to let every customer know that before we let you sign the contract, we're going to give you a statement of work that describes how long it's going to take, and the resources needed on your side, so that everything is transparent." Then the engineering team said they'd look for extra onboarding steps to eliminate. Then our customer support team said, "What if we build a checklist based on positive and negative feedback from companies right after they deploy? We know what the issues are, so let's go back and make sure that we solve for them in advance."

I have a story like that for every department in our company. If one department is having a problem, the answers to that problem are invariably scattered throughout the rest of the organization.

OPERATIONALIZING PADRE

PADRE permeates everything we do at Zuora. We build all our metrics around it, which we share with the whole company: pipeline coverage, sales numbers, new logos deployed, retention statistics, expansion rates. We teach it to all our new hires as part of the onboarding process. We build our operating cadence around it— weekly, quarterly, and annually. Every week, all the managers get an eight-slide deck (one for each subsystem) that shows where we are, with red/green/yellow flags to indicate hot spots. And as we've grown and started segmenting our customers, we've created "Franchise" PADRE reports by geography and customer size.

Once a quarter, as a management team, we'll drill into one of the eight subsystems. With thirteen weeks in a quarter, that gives us ample time to make sure the systems are still humming. Sometimes, if needed, we'll drill into one of the subsystems twice a quarter. At any given time, we're probably retooling one of the systems, the one that seems to be breaking because we've hit some inflection point in the business, often due to scale. But then it's a true cross-functional effort, starting with me and then through to my functional leaders to the specific department heads. So, for example, when we shifted from a single-product company to a multiproduct company, we did major work on the product subsystem, but every department was involved.

PADRE is an operating framework, but perhaps more important, it's a way of formalizing a culture. A culture built around always keeping the same intuitive customer insight that we enjoyed as a small company. A culture built around staying sharp, which becomes more difficult as you get bigger. It's a way of making sure that all our institutional knowledge doesn't become scattered and diffuse—locked up in the heads of just a handful of people. Because at the end of the day your customer insight defines who you are; it's your competitive advantage. Lose the insight, and you lose yourself.

A NEW WORLD OF HAPPY BUSINESS

Once upon a time, we used to know the people we bought from—the butcher, the baker, the blacksmith, the farmer. We used to know the people we sold to, the neighbors in our village. All that knowledge got lost a long time ago, when the Industrial Revolution ushered in the product era. But it's coming back in a big way. And in the process, we're getting rid of a lot of stuff we never really needed in the first place—planned obsolescence, the landfill economy, the whole concept of ownership itself.

We've all been in businesses and organizations that felt slow, heavy. But when you get this model working, you find growth. And while growth brings its own challenges, you'll notice that when you're really growing, everything starts to feel much, much lighter. When that wheel is spinning, you're experimenting with new ideas, you're looking outward instead of inward, and you're learning from insights and not mistakes (well, maybe a few mistakes). But perhaps most important, you actually know who you're doing business with. Everything feels fluid but also cohesive, informed as well as inspired. It's that "digital transformation" experience that everyone is always talking about.

It's also a much happier business. Why? Because subscriptions are the only business model that is *entirely based on the happiness of your customers*. Think about it—when your customers are happy, then they're using more of your service, and telling their friends, and you're growing. You get to start every quarter with predictable revenue. You get to make smart, data-driven decisions. You get to benefit from your own customer insights, which are a huge competitive advantage. We call this the world of happy business: happy customers, with happy companies, reinforcing one another, iterating forever, with no beginning and no end.

ADDENDUM

THE SUBSCRIPTION ECONOMY INDEX

INTRODUCTION

The Subscription Economy Index (SEI) is based on anonymized, aggregated, system-generated activity on the Zuora service, a comprehensive billing and finance platform for subscription-based businesses. It reflects the growth metrics of hundreds of companies around the world and spans a number of industries, including SaaS, media, telecommunications, and corporate services.

The breadth and depth of the billing reflected in this study speak to the rapid ascent of the Subscription Economy. As I've mentioned previously, by 2020, we should expect to see more than 80 percent of software providers shift to subscription-based business models (according to Gartner) and 50 percent of the world's largest enterprises depend on digitally enhanced products, services, and experiences.

Recurring revenue-based business models are not new, but they have exploded in recent years owing to cloud-enabled, pay-as-you-go services. As globalization has placed increasing margin strains on manufacturing and product sales, subscription-based businesses have benefited from stable and predictable revenue projections,

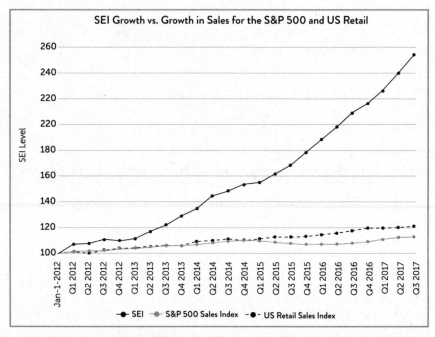

SEI Growth vs. Growth in Sales for the S&P 500 and US Retail

Quarterly levels of the SEI, in comparison to indices of the S&P 500 Sales per Share and US Retail Sales. All indices take a base value of 100 on January 1, 2012, and grow in proportion to the quarterly increase in the one-year trailing total sales that they measure. Over a period of just under six years (January 1, 2012, to September 30, 2017), the SEI grew at an average annual rate of 17.6%. The S&P 500 Sales grows at an average annual rate of 2.2%, while US Retail Sales grew at an average annual rate of 3.6%.

data-driven insights from direct consumer relationships, and large economies of scale owing to relatively small fixed costs.

This study was conducted by Zuora chief data scientist Carl Gold.

Subscription business sales have grown substantially faster than two key public benchmarks—S&P 500 sales and US retail sales. Overall, the SEI reveals that subscription businesses grew revenues about eight times faster than S&P 500 company revenues (17.6 percent versus 2.2 percent) and about five times faster than US retail sales (17.6 percent versus 3.6 percent) from January 1, 2012, to September 30, 2017.

There is a correlation between SEI growth and GDP growth. Both the SEI and GDP slowed around the end of 2016 and begin-

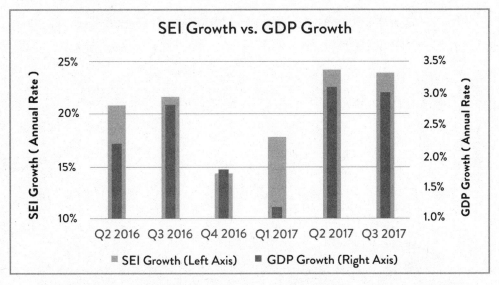

A comparison of SEI growth (left axis) versus American GDP growth (right axis). Note that the SEI generally tracked with the overall GDP slowdown around the end of 2016, and the subsequent acceleration in 2017.

ning of 2017: US GDP growth peaked in Q3 2016 at 2.8 percent and then sank to just 1.2 percent in Q1 2017. At the same time, the SEI growth rate also peaked in Q3 2016 at 21.6 percent, and then cooled to an average annual growth rate of 14.3 percent. Recently, both the SEI and GDP roared back in Q2 and Q3: SEI grew at approximately a 24 percent annual rate for two consecutive quarters, the fastest growth since Q2 2014. At the same time, GDP posted a very strong rebound, growing at a surprise annual rate of 3.1 percent in Q2, the best result since Q1 2015, and a healthy 2.5 percent in Q3.

TWO SUBSCRIPTION ECONOMY GROWTH LEVERS: ARPA AND NET ACCOUNTS

The following figure demonstrates the two primary levers of growth in the Subscription Economy—average revenue per ac-

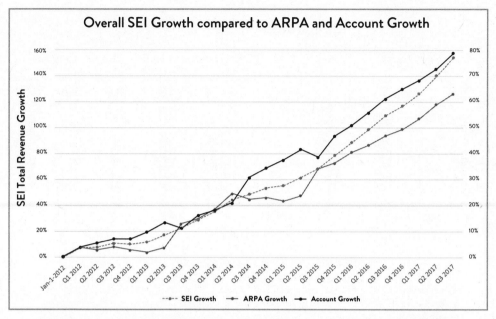

Overall SEI Growth compared to ARPA and Account Growth

Recurring revenue grows through either charging subscribers more (average revenue per account, or ARPA) or charging more subscribers (accounts). The left axis shows cumulative growth of the SEI in percentage terms. The right axis shows the cumulative percentage changes in ARPA and Accounts respectively, both scaled on the right axis. Accounts have grown more or less continuously over the measurement period, while there have been times when ARPA growth slowed and even reversed.

count (ARPA) and net account growth. If the total billings number of a company goes up, that means at least one of two things must have happened—either the number of accounts being billed went up or the amount each account was billed went up.

Note that while the SEI has grown more or less continuously over the last five years, there have been periods when ARPA growth has slowed, and even reversed. There were two discrete periods when companies prioritized net account growth ahead of ARPA growth: 2012–13, and late 2014 to mid-2015. At these times, the total number of accounts grew rapidly, but revenue per account stagnated or sank.

Each of those periods was followed by a correctional phase when the net new accounts decreased, but the average revenue per account increased. Pricing in the Subscription Economy is a flexible, iterative process. Companies frequently experiment with a combination of set fees and usage-based models as they seek to "land and expand." Strategies prioritizing net new account growth will frequently drive growth with competitive pricing, and then later "switch levers" and attempt to drive ARPA with usage-based billing and by upselling into larger accounts.

Note that the most recent two years, 2015–17, appear to represent a "Goldilocks" period of both high net account growth and solid ARPA growth.

SUBSCRIPTION REVENUE GROWTH BY BUSINESS MODEL

The following figure shows the relative growth of B2B, B2C, and B2A (Any) business model subindices. Each subindex "branches" from the primary SEI as it achieves statistical significance relative to the overall data set, which we define as a minimum of twenty-five constituents. By starting each new subindex branch with the value of the base index, they become easier to compare.

Much like basic cohort analysis, this is an effort to display how the past affects the present. For example, while the B2B subindex has experienced the sharpest growth trajectory in recent quarters, it is still clearly recovering from the trough period of 2012–13. Both ARPA and net new accounts were depressed for B2B companies during this period of corporate retrenchment.

For B2B companies, growth rate is the leading indicator of a company's success. In the software sector, for example, a company that grows less than 20 percent annually has a 92 percent chance of failure (McKinsey). Successful B2B companies must scale sales teams, add new product editions and upsell paths, pursue interna-

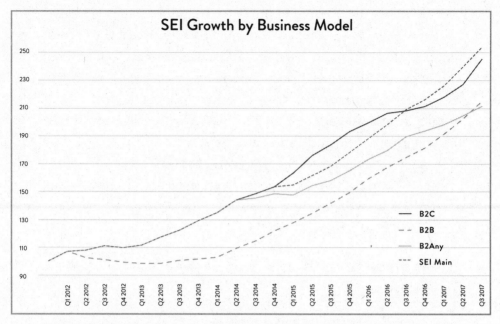

SEI subindices for business models B2B, B2C, and B2A (Any). B2A companies offer subscriptions to both individuals and businesses. Each subindex is launched starting from the value of the main SEI when twenty-five constituents are available. B2B slumped shortly after inception in 2012–13, but has been the fastest-growing subindex in the most recent quarters. B2C was the fastest in 2015, but slowed in 2016. B2A companies generally tracked the main (average) index until the last year when they surprisingly slowed; for now both B2B and B2C are outpacing B2A.

tional markets and larger enterprise accounts, and optimize their business models by taking on usage-based pricing. Their biggest challenges include systems constraints and conflicting systems of record.

For B2C companies, net user growth is the key metric. Successful B2C companies increase subscriber acquisition rates with rapid pricing experimentation, increase retention and ARPA by tailoring offerings based on behavioral insights and willingness to pay, and increase capture rates by taming the complexity of electronic payments. Their challenges include relatively high churn rates owing to poor pricing and packaging decisions, fickle consumer behavior, and/or lost revenue owing to poor payment and acquisition systems. The B2C companies in our study had the fastest growth rate in 2015, but have tapered more recently.

Over the past year (ending in March 2017), B2B and B2C companies in the SEI have had growth rates of 23 percent and 18 percent, respectively. B2B companies have been relatively consistent, while B2C cooled during the late 2016 to early 2017 slowdown. More recently B2C rebounded, posting 7.9 percent (31.8 percent annual rate) growth in Q3. B2A companies generally tracked the main (average) index until the last year, when they surprisingly slowed; for now both B2B and B2C are outpacing B2Any.

Growth rates (last twelve months):

- B2B growth: 23%
- B2C growth: 18%
- B2A growth: 11%

SUBSCRIPTION REVENUE GROWTH BY INDUSTRY

Which industries are thriving in the Subscription Economy? As a subscription billing and finance SaaS company based in Silicon Valley, Zuora has a significant customer base of other software vendors—both SaaS natives and on-premise vendors switching to recurring revenue models. As a result, SaaS was the first subindex—it underperformed the main index in 2013–14, but recently caught up and exceeded the main index. During the slowdown around the end of 2016 in US GDP and SEI growth, SaaS companies were relatively unaffected, while media, telecom, and corporate services all stalled. More recently, media and telecom recovered during the last six months' rebound, but corporate services has remained sluggish, with 4.4 percent growth over the last year.

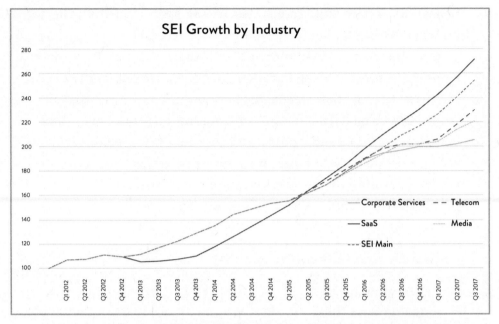

SEI Growth by Industry

SEI subindices for Corporate Services, Telecommunications, SaaS, and Media. Each subindex is launched starting from the value of the main SEI when twenty-five constituents are available. Note that not every SEI constituent falls into one of these categories. SaaS was the first subindex; it underperformed the main index in 2013–14 but recently caught up and exceeded the main index. Recently more subindices have launched and most are close to the performance of the main index.

As Zuora's client base has expanded and diversified, more subindices have launched. Most of these new subindices perform close to the performance of the main index. Early evidence suggests that corporate services are underperforming against the median growth rate, but later editions of this study will confirm this.

Growth rates (last twelve months):

- SaaS growth: 23%
- Telecommunications growth: 14%
- Media growth: 9%
- Corporate Services growth: 4.4%

SUBSCRIPTION REVENUE GROWTH BY REVENUE BAND

Size matters in the Subscription Economy. The subindex made up of $100M+ constituents has been the highest performing since its inception in 2014. In contrast to start-ups, these larger companies have more resources, more distribution, more new acquisitions, more channels to grow. As a result, they benefit from the network effects mentioned earlier in this study.

For start-ups, the real challenges appear to lie after the initial submillion-dollar "honeymoon" growth period. As seen in the figure below, the $1 million to $20 million revenue band has the most challenging growth rate. After the first product has been defined

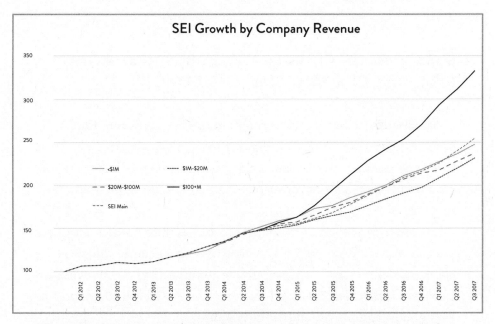

SEI subindices for company size, as given by revenue band. Each subindex is launched starting from the value of the main SEI when twenty-five constituents are available. Note that the revenue band refers to the overall revenue of the constituent company, and not to the specific products hosted on the Zuora platform. The subindex made up of $100M+ constituents has been the highest performing since its inception in 2014.

and initial funding has been received, this is a time when most companies are defining the actual size of their market, which can vary widely. According to McKinsey, only 28 percent of internet services companies reach $100 million in revenue.

Over the last six months, the largest companies grew at an accelerated rate compared with the prior six months, while smaller companies all saw their growth rates decline slightly.

Average annual growth rates (last twelve months):

- < $1M: 17%
- $1M–$20M: 21%
- $20M–$100M: 15%
- $100M+: 31%

SUBSCRIPTION CHURN RATES BY BUSINESS MODEL, INDUSTRY, COMPANY SIZE, AND REGION

At its most basic level, churn refers to the proportion of total subscribers who leave during a given time period. Churn can result from any number of reasons: weak customer service, a poorly upgraded product, a better offer from the competition, business failure, etc.

In order for revenue to recur, customers must renew at a rate that outpaces churn, which can effectively determine the size of a company. Therefore, reducing churn by investing in high-quality services, sticky features, and customer success is fundamental to every subscription-based business strategy.

In addition, reducing churn rates is an imperative not only because of the initial lost revenue, but because of cohort opportunity

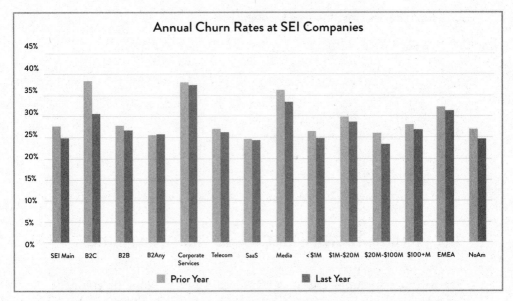

Annual Churn Rates at SEI Companies

Comparison of average annualized churn rates from the SEI subindices for the last year (September 30, 2016–September 30, 2017) and the long-term average over prior years. Overall, churn has decreased in the past year, particularly for Consumer subscription products (B2C) and Media companies (both of which have seen higher average churn rates over the long run). Among the churns in each category, it is highest for B2C, Corporate Services (with Media a close second), $1M–$20M, and in EMEA.

costs—successful accounts grow larger over time. Unsurprisingly, churn rates are higher for B2C and lower for B2B. Digital B2C companies (including media) have large numbers of individual users who frequently churn due to payment challenges, credit card issues, lapsed interest, or competition. B2B companies (in the SEI weighted more heavily toward software) benefit as their solutions become more embedded into stable, growing corporate accounts.

Average annual churn rates in the SEI are generally between 20 and 30 percent. Among the business models, churn is highest for B2C and lowest for B2B companies. For industries, churn is highest in media and lowest in SaaS. Over the past year, churn has fallen overall, particularly for B2C and media companies.

Churn rates (last twelve months):

- B2B: 27%
- B2C: 30%
- B2A: 26%
- Corporate Services: 37%
- Telecommunications: 26%
- SaaS: 24%
- Media: 33%

GROWTH BY REGION: EMEA AND NORTH AMERICA

The SEI now includes EMEA- and North America–specific subindices, beginning in Q1 2017 with one year of history dating back to Q1

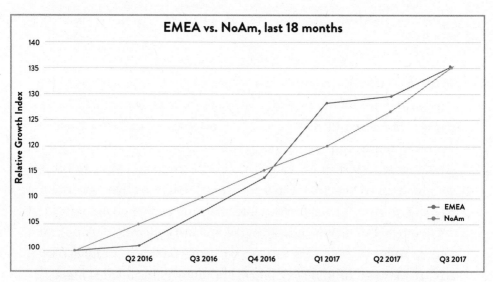

The figure shows the relative growth in recurring revenue for the EMEA Subindex of the SEI and the North America Subindex of the SEI, starting from a base value of 100 at the end of Q1 2016. In each quarter the index is increased by the same percentage as the percentage growth in each region. North American growth started high and slowed somewhat over the course of the year, while EMEA growth was initially slow but completely closed the gap in the second part of the year.

2016. In the last year and a half EMEA and North America indices grew almost exactly the same amount: Since April 2016 EMEA grew a cumulative 35.2 percent (annual rate of 22.3 percent), just beating North America with 34.7 percent cumulative growth (22 percent annual rate). However, growth in the EMEA index has been less consistent, starting slowly in Q2 2016 and having a surprise jump in Q1 2017. But the EMEA index does contain many fewer companies than the North American index does, so this variation probably just reflects noise due to the small sample size—overall it appears that EMEA and North American companies in the Subscription Economy grow at about the same rate.

In short, the Subscription Economy in Europe is clearly on the ascent. Over the past eighteen months, European subscription companies (a new SEI category) have even bested their American counterparts' growth rate of around 22 percent. This is remarkable because European economic growth rates overall have lagged US growth rates for much of the past decade.

SEI UPDATE: A GROWTH GUIDE TO USAGE-BASED BILLING

At its heart, usage-based billing is a way of quantifying value. The goal is to let customers pay for the value they need. The best pricing strategy will let you put a number on the metric that customers value most, based on how they actually use your service. This is commonly called a "value metric." Simply put, a value metric should do three things: align to customer needs, grow with customers, and be predictable (both for the customers and the organization).

According to McKinsey, more than 75 percent of customers want pricing metrics that are aligned with perceived value, easy to understand, and easy to track (and thus predict costs). And yet,

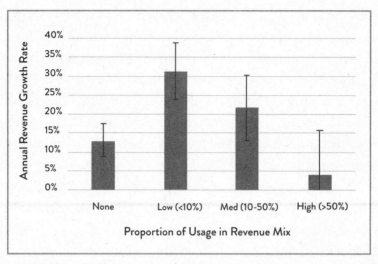

Average growth rate versus proportion of usage billing in the revenue mix: average and 95% confidence error bars. Companies using a small amount of usage-based billing (less than 10%) grew more than 2x faster on average than those companies that use no usage-based billing at all—an average annual growth rate of 31% compared with an average annual growth rate of 13%. Companies using a medium level of usage billing (between 10% and 50%) had an intermediate growth rate of 22%, but the small number of companies with a high proportion of usage billing (more than 50%) had an average growth rate of just 4%; however, the small number of companies in this group means the growth rate for high usage revenue companies lacks statistical significance.

only about 27 percent of subscription businesses use some sort of usage-based pricing today. This is a mistake. Usage models give companies many levers to drive engagement and customer value. Depending on the business, these levers could be the number of seats, emails, API calls, revenue volume, or customer events.

In this latest edition of the SEI, we investigated how much the fastest-growing companies in the Subscription Economy employ usage-based billing in their growth strategies. To do this we compared the annual growth rate of the SEI constituents over a multi-year period, including a total of 550 one-year observations of usage billing and growth metrics. The surprising result is that companies that use just a small amount of usage-based billing in their revenue mix (less than 10 percent) *grew more than twice as fast on average as companies that did not employ usage-based billing at all.*

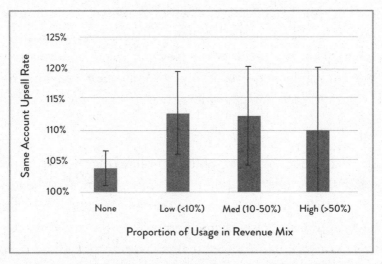

Average same account upsell rate versus proportion of usage-based billing in the revenue mix. Average and 95% confidence error bars. Companies that employ usage-based billing have more than 3x higher upsell rates than companies that do not—around 13% in comparison to 4%.

While a small amount of usage-based billing was associated with a faster growth rate, this association does not necessarily apply to businesses whose billing is majority usage-based. Only a small number of companies have more than 50 percent of their revenue come from usage billing, so that sample size was not large enough to make a strong conclusion; but the evidence suggests these companies may have a lower-than-average growth rate. A third cohort of companies with a medium amount of usage-based billing (between 10 percent and 50 percent of revenue) has an average growth rate that is higher than companies with no usage, but lower than those companies with a small amount of usage-based revenue.

To understand why companies employing usage-based billing grow faster, we examined the factors that drive recurring revenue growth, starting with ARPA growth. When we compared the same account upsell rate on renewals, we found that, across the board, companies employing usage-based billing had higher upsell rates

(12 to 13 percent), in comparison with companies that did not employ usage-based billing (which had an average upsell rate of only 4 percent). This is a result of the fact that usage is normally sold in tiers and tied to the recurring revenue plan, marking out a clear upgrade path for customers who are successfully using a service.

We also examined the second pillar of recurring revenue growth: growth in the number of accounts (or more specifically, the loss of accounts due to churn). Companies that employ usage-based billing have significantly lower churn rates than those that do not, at all levels of usage-based revenue mix. The churn rate for companies employing usage-based billing is about 10 percent less on an annual basis: 26 percent for usage-based billers in comparison with 37 percent for non-usage-based billing companies. These lower churn rates reflect higher customer satisfaction and engagement with companies that fulfill the central tenet of the Subscription Economy: customers get to pay for only what they use.

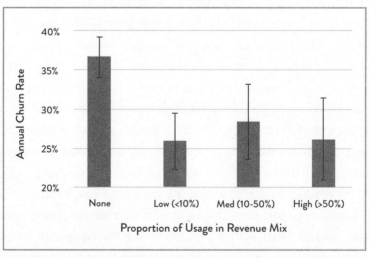

Average churn rate versus proportion of usage-based billing in the revenue mix. Average and 95% confidence error bars.

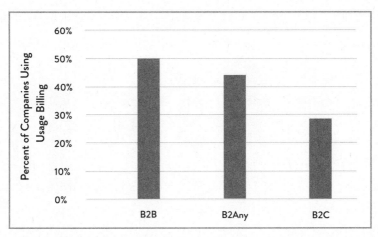

Usage-based Billing by Business Model

To understand which customers are deploying usage-based billing, we looked at adoption by business model and vertical. Usage-based billing is most heavily used by B2B companies, and least used by companies selling direct to consumers: around 50 percent of B2B companies employed usage-based billing, and, of those, the average proportion of usage revenue was more than 25 percent. For companies selling to consumers, less than 30 percent

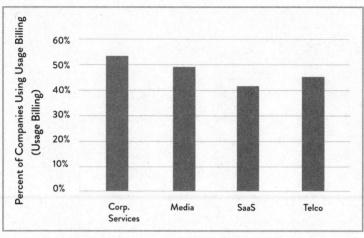

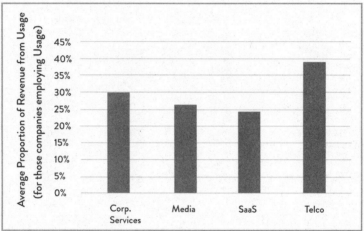

Usage-based Billing by Vertical

employ usage-based billing, and, of those, it accounts for an average of just 16 percent of revenue. Turning to verticals, the deployment of usage-based billing is more consistent, but we were surprised to find that SaaS companies show the lowest deployment of usage-based billing at 41 percent.

Is usage-based billing a surefire growth lever for every company in the Subscription Economy? The track record of top SEI companies suggests huge potential for companies that have not adopted this form of billing, but we advise caution and taking an incremental approach. Service features billed by usage are usually the core value metric of a subscription service offering: the most valuable feature to use, and the most expensive to deliver. If a company charges for use of something that is not clearly providing value, there is the potential to enrage rather than engage!

SEI CONCLUSION

Recurring revenue–based businesses in the Subscription Economy are not guaranteed success, but if they focus relentlessly on extending average customer lifetimes by maximizing ARPA and net account growth, while minimizing churn rates and taking advantage of usage-based billing, they are more likely to achieve the same or faster growth rates covered in this study.

SUBSCRIPTION ECONOMY INDEX METHODOLOGY

The Subscription Economy Index (SEI) tracks the organic growth of subscription businesses by reflecting aggregated, anonymized, system-generated activity on the Zuora subscription management service between January 1, 2012, and September 30, 2017. Zuora updates this study on a biannual basis. To measure average organic recurring revenue growth, Zuora used a weighted average of system activity of more than 365 constituents who had been live on the Zuora platform for at least two years. Because the SEI growth rate is computed as a weighted average, it reflects the typical growth rate of Zuora's customers' revenue, not the rate at which Zuora's customer base is growing. Subindices of the SEI follow the

same exclusions and must include a minimum of twenty-five constituents.*

Sources

S&P Dow Jones Indices
http://us.spindices.com/indices/equity/sp-500

US Census Bureau, "Monthly Retail Trade and Food Services"
www.census.gov/econ/currentdata/dbsearch?program=MRTS&
startYear=1992&endYear=2016&categories=44000&dataType=
SM&geoLevel=US&adjusted=1&notAdjusted=1&submit=GET+
DATA&releaseScheduleId=

McKinsey, "Grow Fast or Die Slow"
www.mckinsey.com/industries/high-tech/our-insights
/grow-fast-or-die-slow

US GDP Growth
www.bea.gov/newsreleases/national/gdp/gdpnewsrelease
.htm

ACKNOWLEDGMENTS

I'd like to start by thanking my cofounders K. V. Rao and Cheng Zuo for taking me on the journey to create the Subscription Economy. I'm grateful to Marc Benioff for giving me the opportunity at Salesforce.com and encouraging me to take my own plunge. Frank Ernst, Ming Li, Monika Saha, Leo Liu, Travis Huch, Xi Yang, Richard Terry-Lloyd, Jeff Yoshimura, Megan Golden, Madhu Rao, and countless other early ZEOs were instrumental to the foundational success of Zuora. Thanks to Tyler Sloat and his phenomenal finance group for changing the way we think about business. Thanks to Mark Heller for deftly managing the logistics of this project, and my assistant Ninni Sonberg for her diligence and patience. Thanks to Jing Jing Xia, An Ly, Todd Pearson, Alvina Antar, Nathan Heller, Steve Yeager, Kevin Suer, and Robert Hildenbrand for their time in helping research the book. Thanks to Erika Malzberg and Aarthi Rayapura for their editorial support and contributions, and Carl Gold for his invaluable data insights. For their design expertise and supervision I'd like to thank Shaun Middlebusher, Lauren Glish, and Peishan Li. Thanks to Jennifer Pileggi and Andy Hackbert for their tremendous legal minds. Thanks to the incredible ecomm group and my friend and partner Marc Diouane.

Our *Subscribed* conference, magazine, and podcast are a constant source of inspiration and insight. In particular I'd like to thank our guests David Wadhwani of AppDynamics, Matt Anderson of Arrow Electronics, Jeff Potter and Mac Kern of Surf Air, Andy Main of Deloitte Digital, Gytis Barzdukas of GE Digital, Anthony Fletcher of Graze, Jamie Allison of Ford, J. B. Wood of the Technology Services Industry Association, Reid DeMarcus of Crunchyroll, Andy Mooney and Ethan Kaplan of Fender, and Sam Jennings, Peter Kreisky, Judy Loehr, Anne Janzer, and Robbie Kellman Baxter.

This has been a singular experience for a first-time author. Thanks to my agent, Jim Levine, at Levine Greenberg, my editor, Kaushik Viswanath, at Penguin Random House and the inestimable Carlye Adler for her feedback and encouragement. Thanks to Jayne Scuncio for her tireless efforts to get our message out there. Thanks to Gabe Weisert for tying himself to the mast.

NOTES

INTRODUCTION

1 *advising people not to go to business school* Tien Tzuo, "Why This CEO Believes an MBA Is Worthless," *Fortune*, April 27, 2015, http://fortune.com/2015/04/27/tien-tzuo-starting-your-own-business.

CHAPTER 1: THE END OF AN ERA

11 *in the year 2000 are now gone* World Economic Forum, "Digital Disruption Has Only Just Begun," Pierre Nanterme, January 17, 2016, www.weforum.org/agenda/2016/01/digital-disruption-has-only-just-begun.

11 *the Fortune 500 first* http://fortune.com/fortune500.

12 *"The digital company. That's also an industrial company"* General Electric television ad, www.ispot.tv/ad/AVhu/general-electric-whats-the-matter-with-owen-hammer.

13 *Disney, ABC, ESPN, Pixar* Elaine Low, "Disney Ditches Twitter, but Does Distribution Talk Point to Netflix?" *Investor's Business Daily*, October 6, 2016, www.investors.com/news/disney-may-be-out-on-twitter-but-its-mulling-distribution-plans.

17 *Age of the Customer* Forrester Research, "Age of the Customer," https://go.forrester.com/age-of-the-customer.

21 *new improvements in digital user experiences* Kleiner Perkins Caufield Byers, "Internet Trends 2017," www.kpcb.com/internet-trends.

CHAPTER 2: FLIPPING THE RETAIL SCRIPT

24 *Consumers already know what to expect* Kantar Retail IQ, "2017 U.S. Retail Year in Review," www.kantarretailiq.com.

24 *top fifteen ecommerce marketplaces* Michael Wolf, "Activate Tech & Media Outlook 2018," http://activate.com.

26 *Smartphone battleground is shifting* "Goldman Says Apple Needs Amazon Prime Subscription Plan," *Bloomberg*, October 17, 2016,

www.bloomberg.com/news/videos/2016-10-17/why-apple
-needs-a-subscription-plan.

29 **blocking the exit** Robbie Kellman Baxter, *The Membership Economy:
Find Your Super Users, Master the Forever Transaction, and Build
Recurring Revenue* (New York: McGraw-Hill Education, 2015).

31 **sell people guitars and then hope they play it** Dan Mullen,
"Fender: Reinventing Guitar for a Digital Age," *Subscribed
Magazine,* September 12, 2017, www.zuora.com/2017/09/12/fender
-reinventing-guitar-for-the-digital-age.

32 **data-driven, app-centric, flexible and omnichannel retailing** Mike
Elgan, "The 'Retail Apocalypse' Is a Myth," *Computerworld*, October
21, 2017, www.computerworld.com/article/3234567
/it-industry/the-retail-apocalypse-is-a-myth.html.

CHAPTER 3: THE NEW GOLDEN AGE OF MEDIA

43 **It's like a beautiful blend of all of them** Gabe Weisert, "Lessons
from New Media: Crunchyroll Conquers the World," *Subscribed
Magazine,* May 12, 2016, www.zuora.com/2016/05/12/lessons-from
-new-media-crunchyroll-conquers-the-world.

44 **CEO James Rushton told Sportsmail** Amitai Winehouse, "How the
'Netflix of Sport' Could Change the Way Supporters Watch
Football," *Mail Online*, September 5, 2017, www.dailymail.co.uk
/sport/football/article-4854722/Behind-DAZN-New-Netflix-sport
-changing-watch.html.

45 **analyst Craig Moffett told Recode** Peter Kafka, "Another Half-
million Americans Cut the Cord Last Quarter," *Recode*, May 3, 2017,
www.recode.net/2017/5/3/15533136/cord-cutting-q1-half
-million-tv-moffett.

47 **subscription model of buying music is bankrupt** Kevin Fogarty,
"Tech Predictions Gone Wrong," *Computerworld*, October 22, 2016,
www.computerworld.com/article/2492617/it-management/tech
-predictions-gone-wrong.html.

47 **"Music is going to become like running water"** John Paul Titlow,
"David Bowie Predicted the Future of Music in 2002," *Fast
Company,* January 11, 2016, www.fastcompany.com/3055340/david
-bowie-predicted-the-future-of-music-in-2002.

48 **more like members, and less like customers** "Subscribed Podcast
#6: Sam Jennings on Prince and the Music Streaming Business,"
www.zuora.com/guides/subscribed-podcast-ep-6-sam-jennings
-prince-music-streaming-business.

50 **NPG Music Club 4ever** "When Did NPG Music Club Start and
Finish?" http://prince.org/msg/7/349218.

CHAPTER 4: PLANES, TRAINS, AND AUTOMOBILES

51 *all those steps are very complicated, particularly when you talk to millennials* Nick Lucchesi, "We Are Entering the Era of Car Subscriptions," *Inverse*, November 18, 2016, www.inverse.com /article/24012-hyundai-ioniq-subscription.

52 *Netflix or Apple's iPhone [upgrade] program* Nick Kurczweski, "Buy, Lease or Subscribe? Automakers Offer New Approaches to Car Ownership," *Consumer Reports*, October 11, 2017, www .consumerreports.org/buying-a-car/buy-lease-or-subscribe -automakers-offer-new-approaches-to-car-ownership.

52 *get a car subscription for two months* Christina Bonnington, "You Will No Longer Lease a Car. You Will Subscribe to It,"*Slate*, December 2, 2017, www.slate.com/articles/technology /technology/2017/12/car_subscriptions_ford_volvo_porsche_and _cadillac_offer_lease_alternative.html.

55 *a sports bar, with a bartender* "The Rev-Up: Imagining a 20% Self Driving World," *The New York Times*, November 8, 2017, www .nytimes.com/interactive/2017/11/08/magazine/tech-design-future -autonomous-car-20-percent-sex-death-liability.html?_r=0.

55 *250 million connected cars on the road by 2020* "Gartner Says by 2020, a Quarter Billion Connected Vehicles Will Enable New In-Vehicle Services and Automated Driving Capabilities," January 26, 2015, www.gartner.com/newsroom/id/2970017.

56 *Without control over the platform, PC hardware* Horace Dediu, "IBM and Apple: Catharsis," July 15, 2014, www.asymco.com/2014 /07/15/catharsis.

58 *the cars weren't connected* "Subscribed San Francisco 2017 Opening Keynote," Zuora *Subscribed* conference presentation, June 5, 2017, www.youtube.com/watch?v=fdDA7sRgMSQ.

63 *Americans aged 20–24 with a driving licence* "Transport as a Service: It Starts with a Single App," *The Economist*, September 29, 2016, www.economist.com/news/international/21707952 -combining-old-and-new-ways-getting-around-will-transform -transportand-cities-too-it.

CHAPTER 5: COMPANIES FORMERLY KNOWN AS NEWSPAPERS

66 *evidence of diminishment* Eric Alterman, "Out of Print," *The New Yorker*, March 31, 2008, www.newyorker.com/magazine/2008/03/31 /out-of-print.

67 *use ad blockers on a daily basis* Reuters Institute 2017 Digital News Report, www.digitalnewsreport.org.

68 *die or be forced to find another way* Josh Marshall, "There's a Digital Media Crash. But No One Will Say It," *Talking Points Memo*, November 17, 2017, http://talkingpointsmemo.com/edblog/theres-a -digital-media-crash-but-no-one-will-say-it.

68 *more relevant than in the past* Jessica Lessin, "What Everyone Is Missing About Media Business Models," *The Information*, January 6, 2017, www.theinformation.com/articles/what-everyone-is -missing-about-media-business-models.

69 *most important strategic goal for most news publishers* Ken Doctor, "Newsonomics: The 2016 Media Year by the Numbers," *Newsonomics*, December 19, 2016, www.niemanlab.org/2016/12/newsonomics-the -2016-media-year-by-the-numbers-and-a-look-toward-2017.

72 *roughly analogous to what the magazines used to be* Sahil Patel, "With a Billion Views on YouTube, Motor Trend Is Now Building a Paywall," *Digiday*, February 15, 2016, https://digiday .com/media/nearly-billion-views-youtube-motor-trend-now -building-video-paywall.

74 *increase our renewal subscription revenue* Aarthi Rayapura, interview with Subrata Mukherjee, vice president of product at *The Economist*, *Subscribed Magazine*, June 16, 2016, https://fr.zuora.com /2016/06/16/focusing-subscription-economy-subrata-mukherjee -vp-product-economist.

77 *you need to have a direct connection with them* Lucia Moses, "To Please Subscription-Hungry Publishers, Google Ends First Click Free Policy," *Digiday*, October 2, 2017, https://digiday.com /media/please-subscription-hungry-publishers-google-ends-first -click-free-policy.

78 *sounder business strategy for the* **Times** "Journalism That Stands Apart," The Report of the 2020 Group, *The New York Times*, January 2017, www.nytimes.com/projects/2020-report/.

CHAPTER 6: SWALLOWING THE FISH: LESSONS FROM THE REBIRTH OF TECH

80 *"from a revenue perspective"* David McCann, "Adobe Completes Swift Business Model Transformation," *CFO*, August 18, 2015, ww2 .cfo.com/transformations/2015/08/adobe-completes-swift-business -model-transformation/.

81 *moving people up the product ladder* "The Reinventors: Adobe," *Subscribed* conference presentation.

81 *our revenue dropped about 20 percent* McKinsey & Company, "Reborn in the Cloud," www.mckinsey.com/business-functions /digital-mckinsey/our-insights/reborn-in-the-cloud.

83 *"IT Doesn't Matter"* Nicholas G. Carr, "IT Doesn't Matter," *Harvard Business Review*, May 2003, https://hbr.org/2003/05/it-doesnt-matter.

84 *80 percent of software providers will have shifted* Christy Pettey, "Moving to a Software Subscription Model," www.gartner.com /smarterwithgartner/moving-to-a-software-subscription-model.

84 *consumption-based offerings* Deloitte, "Flexible Consumption Transition Strategies for Business," www2.deloitte.com/us/en /pages/technology-media-and-telecommunications/articles /flexible-consumption-transition-strategies.html.

85 *tumultuous period of costs exceeding revenue* Thomas Lah and J. B. Wood, *Technology-as-a-Service Playbook: How to Grow a Profitable Subscription Business* (Seattle: Point B Inc., 2016).

93 *you don't have to go all in* Jaakko Nurkka, Josef Waltl, and Oliver Alexy, "How Investors React When Companies Announce They're Moving to a SaaS Business Model," *Harvard Business Review*, January 2017, https://hbr.org/2017/01/how-investors-react-when -companies-announce-theyre-moving-to-a-saas-business-model.

95 *tighter relationship with customers* Matt Brown, "Cisco's Software Strategy Is Resonating with Customers, Driving Business," *CRN*, November 28, 2017, www.crn.com/news/networking/300095901 /partners-ciscos-software-strategy-is-resonating-with-customers -driving-business.htm?itc=ticker.

96 *compared with 11 percent growth* Charles S. Gascon and Evan Karson, "Growth in Tech Sector Returns to Glory Days of the 1990s," Federal Reserve Bank of St. Louis, *Regional Economist*, Second Quarter 2017, www.stlouisfed.org/publications/regional-economist/second-quarter -2017/growth-in-tech-sector-returns-to-glory-days-of-the-1990s.

CHAPTER 7: IOT AND THE FALL AND RISE OF MANUFACTURING

98 *big construction projects* Nicklas Garemo, Stefan Matzinger, and Robert Palter, "Megaprojects: The Good, the Bad, and the Better," McKinsey & Company, www.mckinsey.com/industries/capital -projects-and-infrastructure/our-insights/megaprojects-the-good -the-bad-and-the-better.

98 *When the Komatsu team arrives at your job site* "Smart Construction," video, Komatsu America Corporation, www.youtube .com/watch?v=aZdtPhMg3dY.

99 *even before there is a work site* "Tom Bucklar, Caterpillar," Zuora *Subscribed* conference, www.youtube.com/watch?v=Qio2oGJ_G_o.

100 *overall employment in manufacturing in the United States* Bureau of Labor Statistics "Manufacturing: NAICS 31-33," www .bls.gov/iag/tgs/iag31-33.htm.

101 ***drop in global productivity*** IMF Staff Discussion Note, "Gone with the Headwinds: Global Productivity," April 3, 2017, www.imf.org/~ /media/Files/Publications/SDN/2017/sdn1704.ashx.

101 ***manufacturers contributed $2.2 trillion*** National Association of Manufacturers, "Top 20 Facts About Manufacturing," www.nam .org/Newsroom/Facts-About-Manufacturing.

102 ***And if you sell technology to help sense conditions*** Scott Pezza, "How to Make Money with the Internet of Things," Blue Hill Research, May 18, 2015, http://bluehillresearch.com/how-to-make -money-with-the-internet-of-things.

104 ***most of our factories look the same*** Olivier Scalabre, "The Next Manufacturing Revolution Is Here," TED talk, May 2016, www.ted .com/talks/olivier_scalabre_the_next_manufacturing_revolution _is_here/transcript.

104 ***We recently hosted Gytis Barzdukas*** "Gytis Barzdukas, GE Digital," Zuora *Subscribed* conference, www.youtube.com/watch ?v=OEq5HTz7MDE.

109 ***chief digital transformation officer and president Matt Anderson*** Gabe Weisert, "Arrow Electronics: The Biggest IoT Innovator You've Never Heard Of," Zuora *Subscribed Magazine*, www.zuora.com/guides /arrow-electronics-the-biggest-iot-innovator-youve-never-heard-of.

110 ***What's the value I can create*** Guillaumes Vives, "How Do You Price a Connected Device?" Zuora, November 19, 2015, www.zuora .com/2015/11/19/how-do-you-price-a-connected-device.

111 ***Everything that we formerly electrified*** Kevin Kelly, *The Inevitable: Understanding the 12 Technological Forces That Will Shape Our Future* (New York: Viking, 2016).

113 ***This 'as-a-service' approach can give the supplier*** McKinsey & Company, "Unlocking the Potential of the Internet of Things," www.mckinsey.com/business-functions/digital-mckinsey/our-insights /the-internet-of-things-the-value-of-digitizing-the-physical-world.

CHAPTER 8: THE END OF OWNERSHIP

114 ***digitally enhanced products, services, and experiences*** International Data Corporation, "IDC Sees the Dawn of the DX Economy and the Rise of the Digitally Native Enterprise, International Data Corporation," November 1, 2016, www.idc.com /getdoc.jsp?containerId=prUS41888916.

117 ***what if business school consisted of ten months*** Steve Kolowich, "Would Graduate School Work Better If You Never Graduated from It?" *Chronicle of Higher Education*, July 17, 2014, www.chronicle

.com/blogs/wiredcampus/would-graduate-school-work-better-if
-you-never-graduated-from-it/54015.

119 *Not only are renewables playing a far bigger role* "All Change: New
Business Models," *The Economist,* January 15, 2015, www.economist.
com/news/special-report/21639019-power-industrys-main-concern
-has-always-been-supply-now-it-learning-manage.

121 *only 7 percent of bank credit products* Karen Mills and Brayden
McCarthy, "How Banks Can Compete Against an Army of Fintech
Startups," *Harvard Business Review,* April 2017, https://hbr.org/2017
/04/how-banks-can-compete-against-an-army-of-fintech-startups.

CHAPTER 9: THAT WTF MOMENT

127 *you have to think about the long term* Emanuel Maiberg, "Final Fantasy
Producer Says Subscriptions Still Make Sense for MMOs," *GameSpot,*
March 30, 2014, www.gamespot.com/articles/final-fantasy-producer
-says-subscriptions-still-make-sense-for-mmos/1100-6418646.

130 *focusing on systems of innovation* Gartner, "Gartner Says Adopting
a Pace-Layered Application Strategy Can Accelerate Innovation,"
February 14, 2012, www.gartner.com/newsroom/id/1923014.

130 *you have to provide real value and solve problems* Anne Janzer,
"Subscription Marketing: Strategies for Nurturing Customers in a
World of Churn," Cuesta Park Consulting, 2017.

CHAPTER 10: INNOVATION: STAYING IN BETA FOREVER

134 *re-enable the beta label for Gmail* "Google Apps Is Out of Beta
(Yes, Really)," Official Google Blog, July 7, 2009, https://googleblog
.blogspot.com/2009/07/google-apps-is-out-of-beta-yes-really.html.

135 *The Manifesto for Agile Software Development* Manifesto for
Agile Software Development, http://agilemanifesto.org.

139 *a show like* **House of Cards** David Carr, "Giving Viewers What
They Want," *The New York Times,* February 25, 2013, www.nytimes
.com/2013/02/25/business/media/for-house-of-cards-using-big
-data-to-guarantee-its-popularity.html.

140 *cable networks don't know that stuff* Mark Sweney, "Netflix
Gathers Detailed Viewer Data to Guide Its Search for the Next Hit,"
The Guardian, February 23, 2014, www.theguardian.com/media
/2014/feb/23/netflix-viewer-data-house-of-cards.

140 *Howard Schultz said on a Starbucks earnings call* Paul R.
LaMonica, "Starbucks Still Has a Problem with Long Lines," CNN,
January 27, 2017, http://money.cnn.com/2017/01/27/investing
/starbucks-long-lines-mobile-ordering-earnings/index.html.

141 *deliver technology that enhances the human connection* Clint Boulton, "Starbucks' CTO Brews Personalized Experiences," *CIO*, April 1, 2016, www.cio.com/article/3050920/analytics/starbucks -cto-brews-personalized-experiences.html.

CHAPTER 11: MARKETING: RETHINKING THE FOUR P'S

147 *general education about how subscriptions work* Greg Alexander, "How to Transition Channel Partners from Selling Perpetual Licenses to SaaS," Sales Benchmark Index, May 20, 2017, https: //salesbenchmarkindex.com/insights/how-to-transition-channel -partners-from-selling-perpetual-licenses-to-saas.

154 *perfectly respectable entry-level service* Madhavan Ramanujam and Georg Tacke, *Monetizing Innovation: How Smart Companies Design the Product Around the Price* (Hoboken, NJ: Wiley, 2016).

CHAPTER 12: SALES: THE EIGHT NEW GROWTH STRATEGIES

165 *cross-sold multiple services to about one third* Kevin Chao, Michael Kiermaier, Paul Roche, and Nikhil Sane, "Subscription Myth Busters: What It Takes to Shift to a Recurring-Revenue Model for Hardware and Software," McKinsey & Company, December 2017, www.mckinsey.com/industries/high-tech/our -insights/subscription-myth-busters.

168 *users in that organization having adopted Box* Eugene Kim, "After 11 Years, Box's CEO Understands the Best Way to Sell to Big Companies," *Business Insider*, August 20, 2016, www.businessinsider .com/box-ceo-aaron-levie-future-enterprise-sales-2016-8.

173 *it has a 92 percent chance of failure* Eric Kutcher, Olivia Nottebohm, and Kara Sprague, "Grow Fast or Die Slow," McKinsey & Company, www.mckinsey.com/industries/high-tech/our -insights/grow-fast-or-die-slow.

CHAPTER 13: FINANCE: THE NEW BUSINESS MODEL ARCHITECTS

177 **Summary of Arithmetic, Geometry, Proportions and Proportionality** Tim Harford, "Is This the Most Influential Work in the History of Capitalism?" BBC, October 23, 2017, www.bbc .com/news/business-41582244.

184 *We call it Tyler's Slide* Tyler Sloat, "An Introduction to Subscription Finance," Zuora Academy, www.zuora.com/guides /subscription-finance-basics.

INDEX

INDEX